W9-BNO-247

An Introduction to International Telecommunications Law

For a complete listing of the *Artech House Telecommunications Library,*
turn to the back of this book.

An Introduction to International Telecommunications Law

Charles H. Kennedy
M. Veronica Pastor

Artech House
Boston • London

Library of Congress Cataloging-in-Publication Data
Kennedy, Charles H.
An introduction to international telecommunications law / Charles H. Kennedy.
p. cm.
Companion volume to: An introduction to U.S. telecommunications law / Charles H. Kennedy. c1994.
Includes bibliographical references and index.
ISBN 0-89006-835-6 (alk. paper)
1. Telecommunication—Law and legislation. I. Kennedy, Charles H. Introduction to U.S. telecommunications law. II. Title.
K4305.4.K46 1996
343.09'94—dc20
[342.3994]
95-51966
CIP

British Library Cataloguing in Publication Data

Kennedy, Charles H.
An introduction to international telecommunications law
1. Telecommunication—Law and legislation
I. Title
342.3'994

ISBN 0-890068356

685 Canton Street
Norwood, MA 02062

International Standard Book Number: 0-89006-835-6
Library of Congress Catalog Card Number: 95-51966

10 9 8 7 6 5 4 3 2 1

For two fine engineers—Charles H. Kennedy, Jr., and Robert F. Vitek

Charles H. Kennedy

For my sister, Eugenia, and for my parents, María Rivera Delgado and José María Pastor Freixa

M. Veronica Pastor

Contents

Preface

Like its companion volume, *An Introduction to U.S. Telecommunications Law*, this book is intended to give nonspecialists a readable survey of a difficult subject. Like its companion book, it also assumes no prior knowledge of telecommunications law or technology and avoids jargon wherever possible.

We have tried to make this book useful to a wide range of readers, and in doing so we probably have not achieved any individual's ideal. So, for example, attorneys who practice telecommunications law may find the relative paucity of legal citations distracting, and engineers will certainly find our technical descriptions unilluminating; while those who specialize in neither field may think we have indulged too much in legalese, footnotes, and Computerspeak. We are content if each reader, nonetheless, finds something of value here that was not readily available elsewhere.

Each of us owes a debt to those who have helped with this project or tolerated us while we worked on it. We name only a few of them here.

CHK wishes to thank three experts at Bell Atlantic Network Services, Inc., for their helpful comments on portions of the manuscript: Chuck Eppert, Director of Technology Planning; Roger Nucho, Director, Standards; and Harry Hetz, Manager of Advanced Intelligent Network Mediated Access. CHK also wishes to thank Marney, Brendan, and Cassie for their unfailing patience.

MVP gives special thanks to Jaime Palacios, Tycho H. E. Stahl, Colm P. MacKernan, and Rafael Zamora for their help and support, and gives very special thanks to Francis J. Skrobiszewski for reading the manuscript when he had better things to do.

Finally, both of the authors thank Mark Walsh, Kimberly Collignon, Laura Esterman, Darrell Judd, and the rest of the editorial, production, and promotions staff of Artech House for their enthusiastic guidance and support.

Introduction: The Global Reach of Telecommunications

Both the sophistication of telecommunications services and the ease with which those services cross national boundaries have advanced dramatically in recent years. To ordinary telephone and telegraph service we have added data communications, electronic mail, facsimile transmission, video, and wireless telephony. And with the aid of satellites and other wireless technologies, these services are available not only to nations with an extensive communications infrastructure, but also to people living in the remotest and least-developed corners of the earth.

These changes have been accompanied by a rapid evolution of the rules under which communications services are provided. Among nations, commercial, scientific, and political arrangements are adjusting to the need to move information seamlessly around the globe. Within nations, a growing trend toward privatization and competition is accelerating the implementation of new technologies and increasing the number of providers to which consumers can turn. This book surveys the law of international telecommunications from both of these perspectives. It first describes the rules under which information moves from one nation to another and then surveys the rules under which individual nations regulate the movement of information within their borders.

As a guide to what follows, we briefly introduce here the issues posed by each of these subjects.

Regulation of Telecommunications Among Nations

New communications technologies tend to be used first to link users within countries and only later to link users in different countries. Achieving this second stage of development presents legal and regulatory challenges that do not arise—or do not arise with the same force—when transmissions originate and terminate within the borders of a single nation.

One set of issues posed by international communications involves *technical compatibility* of the systems that are linked together to carry international messages. So, for example, if two countries' telephone systems use different signaling methods or use different data transmission protocols to send and receive digitized information, then those systems will not carry messages jointly unless the two countries agree to standardize on one technology or the other, or adopt some method of converting the signal as it moves from one system to the next. So-called standards issues, which are the subject of Chapter 2, have been with us since the earliest days of telecommunications and have grown so complex that they require the full-time attention of international committees employing thousands of people.

Another set of issues, covered in Chapter 3, involves the allocation among nations of scarce resources needed for electronic transmission. The best-known example of such a resource is the electromagnetic radio spectrum, which is allocated primarily by a United Nations organization called the International Telecommunications Union (ITU). The other prominent example is the set of orbital positions suitable for geostationary communications satellites, which are also allocated in part by international agreement. Without multinational cooperation in the use of these resources, satellite and terrestrial radio communications would degenerate into an electronic war of all against all.

International cooperation has also been typified by the development and operation of the International Satellite Communications Organization (INTELSAT) system—the principal provider of satellite communication services in the world—and the coordination of INTELSAT with other, so-called separate satellite systems. The arrangements and rules under which these systems operate are the subjects of Chapters 4 and 5.

The terms and conditions under which international telecommunications services are provided are also affected by the regulations imposed by particular countries and are disproportionately affected by the regulations of the U.S. Federal Communications Commission (FCC). The FCC's approaches to licensing, tariffs, accounting rates, settlements, and other issues affecting international services are described in some detail in Chapters 6 and 7.

Finally, the movement of information across borders is affected by national and multinational rules and policies affecting trade regulation, copyright, and data protection (or electronic privacy). Some issues and current developments in each of these areas are introduced in Chapter 8.

As we move through our survey of the regulation of communications between nations, the issues we have listed here will reappear in many contexts. We shall find that some of these subjects (such as technical standards and allocation of frequencies and orbits) have been the focus of extensive international cooperation involving elaborate institutional arrangements, while others (such as licensing and terms of service) remain largely matters of unilateral regulation by individual nations.

Regulation of Telecommunications Within Nations

Almost since the invention of the telegraph, governments have controlled the provision of electronic communications services. In most countries, the service providers were simply nationalized, often within the existing postal authority. In others (notably the United States), telecommunications services were offered by private monopolies at prices and terms subject to the approval of regulatory commissions. Whichever regime was chosen, the result was to insulate each nation's telecommunications industry from domestic and foreign competitors.

Since the 1970s, the telecommunications industries of the world have undergone a dramatic transition. The change began with the divestiture of the AT&T monopoly in the United States and the liberalization of telecommunications in the United Kingdom, and has grown (slowly at first) into what now appears to be an inexorable, worldwide trend toward privatization and competition.

As this trend toward liberalization of telecommunications regulation continues, the rules under which private investment—both domestic and foreign—can penetrate national telecommunications markets and the choices available to users of telecommunications will continue to vary considerably from one country to the next. In the second part of this book, we describe the issues posed by the liberalization process in general terms (Chapter 9) and summarize the history and status of this transition in Asia, North America, the European Union, Eastern Europe, and Latin America (Chapters 10 through 14).

Conclusion

If a single theme can be said to unite a group of subjects as diverse as those covered in this book, it might be the evolution of telecommunications regulation from a system of state control of market entry and terms of service to a system in which private initiative assumes the principal role. This trend is obvious, of course, in Part II of the book, which describes the movement of domestic telecommunications regimes from state monopoly to privatization and competition; but it is no less true in Part I, where we see international standards moving in the direction of greater openness, new satellite systems eroding the INTELSAT monopoly, and radio spectrum allocations driven by private interests, such as the emerging providers of low-Earth-orbit satellite communication systems. International cooperation in telecommunications is no longer a diplomatic game played exclusively by governments acting on behalf of their domestic monopolies, but is increasingly a means of facilitating a global free market in communications.

Part I:
Telecommunications Among Nations

Telecommunications Services: What They Are and How They Work

1

Telecommunications may be defined as the electronic transmission of *information* chosen by the sender between or among *places* also chosen by the sender. While this sounds simple enough, the definition embraces a universe of very different services and technologies—different in the information they convey, in the equipment and techniques they use to represent and transmit that information, and in the private or governmental entities that provide them.

Before we can make sense of the legal and regulatory problems these services present, we should understand what they do and the technical and business arrangements on which they are based. This chapter takes several of the principal services in turn, beginning with the most familiar face of telecommunications—message telecommunications service (MTS).

1.1 MESSAGE TELECOMMUNICATIONS SERVICE

MTS is the ordinary switched, two-way telephone service that we use to talk with friends, relatives, and business associates, and to send and receive low-speed data and facsimile transmissions in the form of tones generated by a modem. As a business, MTS can be described as an offering of access to a network for the point-to-point transmission of conversations, for which users typically pay through some combination of fixed and metered charges. The service may be provided by a government agency or by a private corporation; end-to-end transmission may be provided by a single carrier, or the service may be divided among two or more providers. MTS is, of course, the most pervasive telecommunications service in the world.

1.1.1 The Technology of MTS

From the user's perspective, few things are simpler and more instinctive than an MTS call; yet few things are more complex than the way this simplicity is achieved. In order to make some sense of this complexity, we might divide the MTS call into two stages: the setup stage and the conversation stage.

Setting Up an MTS Conversation

The call setup process begins when the caller lifts the telephone receiver. A direct current (i.e., a simple flow of electrons) begins in the pair of wires (called an *access line* or *local loop*) connecting the caller's telephone to the nearest telephone company switch. The switch detects the flow of current and sends a dial tone, which signals the user that the switch is ready to receive the called party's telephone number. When the caller sends the number of digits the switch is programmed to expect, the switch reads the number to determine where the call will be sent.

Once it knows where to send the call, the local switch must establish a path for the call to take to its destination. If the call will go to another telephone in the caller's exchange (i.e., within the territory served by the same local switch), the switch will simply connect the caller's access line with the access line of the called party. If the call must go to a distant exchange, the switch will connect the caller to a high-capacity interoffice transmission facility called a *trunk*. Depending on the distance separating the two parties' exchanges, the caller may be linked with a number of switches connected by many trunks—perhaps including toll facilities that handle only long-distance traffic—before the connection with the called party's telephone is complete. When the appropriate series of switches, trunks, and lines has been linked, they form a temporary connection called a *circuit*.

If the called party's telephone is not in use, a ringing signal now alerts the called party to the incoming call. If the party answers, the caller's telephone company will begin timing the call for billing purposes. The circuit is now ready to carry the two-way voice conversation between the parties. With the setup stage complete, the first words spoken by the parties begin the conversation stage.

Transmitting an MTS Conversation

The parties' words begin, like all sound, as a rapid compression and decompression of air caused by the vibration of a body (in this case, the speaker's larynx). These alternate compressions and decompressions travel through the air in a wave like ripples in a pond and encounter the mouthpiece of the telephone handset. Inside the mouthpiece, a substance (usually a diaphragm in contact with a handful of carbon granules) stands between the moving air and the electrical current already flowing between the telephone and the local telephone company switch. The compaction of the granules varies with the air pressure of the speaker's sound wave, and those changes in compaction vary the resistance of the wires to the flow of electricity.

When the speaker says "hello," the flat electrical current, known as *direct current* (dc), flowing in the wire begins to undulate at the same frequency and with the same changes in amplitude as the acoustic wave of the speaker's voice. So if the acoustic wave oscillates at 2,000 Hz, the electrical current will oscillate at 2,000 Hz. The current has become an *analog* of the speech input—correctly suggesting that what we have here is an analog transmission.

When the called party's "hello" reaches the central office, it must be transferred to the trunk facility selected during the call setup process. At this point, in most modern telephone networks, the "hello" will be changed from an analog to a digital signal. This means that the caller's speech waveform will be measured, converted into a set of binary (computer-readable) numbers, and transmitted in pulses rather than as a continuous analog wave.

Figure 1.1 should help to explain the measuring stage of the digitization process. The trick is to measure the amplitude of the analog waveform at different times—a process called *sampling*—and to record and transmit those measurements in digital form. The more frequently these samples are taken, the more accurately the measurements will approximate the actual waveform of the speaker's voice.

The sampling process sounds simple enough, but is complicated by the fact that speech waveforms oscillate thousands of times each second and include enormously complex overtones that give each voice its individual quality. In order to measure these values with acceptable accuracy, the telephone companies sample speech at a standard rate of 8,000 samples per second [1].

The next step is to quantize each sample of the speaker's voice—that is, to assign each sample a numerical value on a scale of magnitude. Some "rounding off" occurs in this process; in the present telephone company digital systems, all magnitudes are rounded to the nearest of 256 levels.

Once we have quantized each sample, we must translate these values into a code suitable for transmission and then represent these coded values electronically.

The telephone companies have solved the first problem by translating the numbers into a binary format—that is, into the format used by computers. For those who are not familiar with this format, we should say a word about how this is done.

A binary number is written as a series of ones and zeros, with each position in the series representing a power of two. As in our decimal system (which is based on powers of ten), the digits ascend in magnitude as we read from right to left. So, for example, if a 1 appears in the extreme right-hand position, it rep-

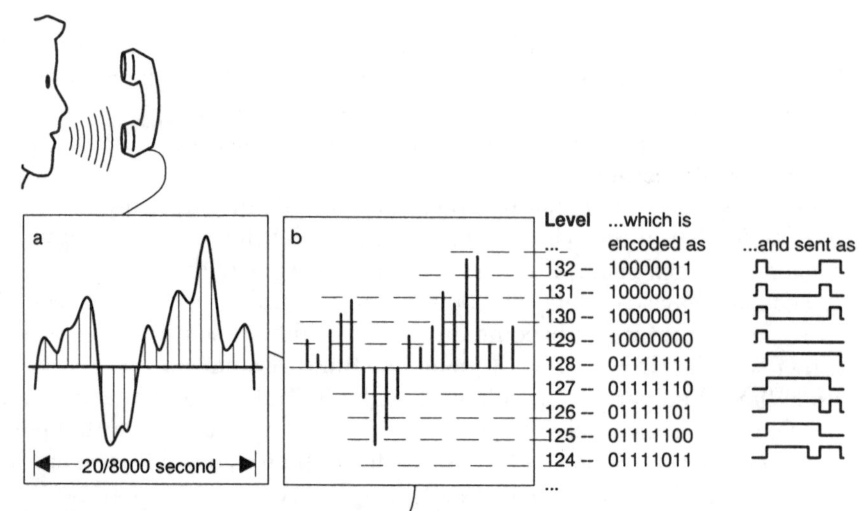

In a digital system, the soundwave is **sampled** (a) at sufficiently close intervals (1/8000 of a second) to very accurately reproduce the wave's shape. The amplitudes of the samples are then **quantized** (b) -- or given approximate values according to the range into which the amplitude falls. The new signal is encoded into an 8-character binary format (which permits 256 possible levels) for transmission through the network. In this example, the digitized signal would be:

10000000,01111111,10000000 ... (129, 128, 129...)

The digital signal is regenerated rather than amplified (as in analog) during transmission; the repeater reads the deteriorating signal (c) and generates a fresh sequence of 1s and 0s (d).

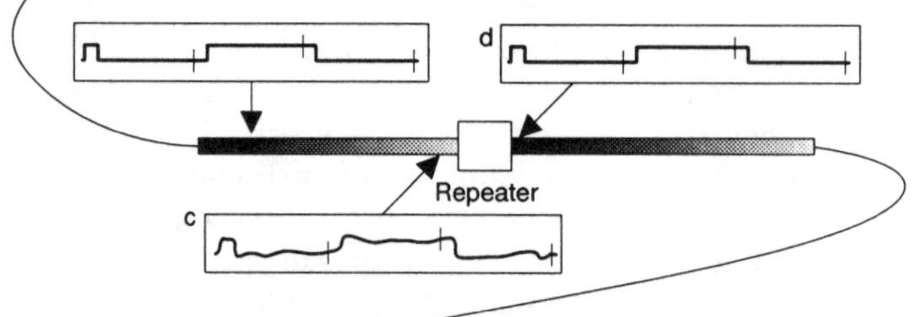

Finally, the signal is converted back into an electical impulse (e, f) and to soundwaves (g).

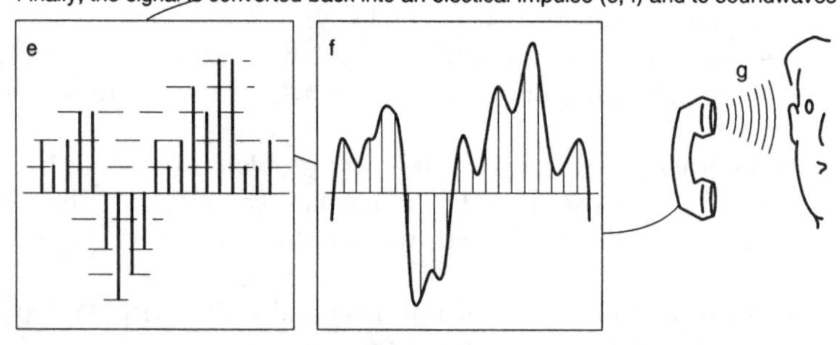

Figure 1.1 Digitization of human speech in a modern telephone network. (*Source:* U.S. Congress, Office of Technology Assessment.)

resents a 1; if a 1 appears in the next position to the left, it represents a 2; if a 1 appears in the third position, it represents a 4. (As in the decimal system, a 0 in any position is a placeholder.) The following are examples of binary numbers: the binary equivalent of the number 1 is 0001; the binary equivalent of the number 2 is 0010; and the binary equivalent of the number 6 is 0110.

Digital carrier systems typically represent each amplitude value of the voice signal as an 8-digit binary number. (The eighth digit may be reserved for signaling information.) So, for example, if a sample of the voice has an amplitude of 100 on the 256-level scale, that amplitude can be translated into the 8-digit binary number 01100100.

This much has given us the *logical* structure of our encoded version of the speech waveform; but how will these binary numbers be represented *electronically*?

The obvious solution is simply to turn the electrical signal off to represent a 0 and turn it back on to represent a 1. This approach has the virtue of simplicity; in order to decide whether it is receiving a 0 or a 1, the receiver has only to detect the presence or absence of current passing through the transmission channel.

This approach is used, in fact, with T-carrier systems (wire pairs used to connect telephone switches over short to moderate distances) and with optical-fiber carrier systems [2]. T-carrier systems are based on the flow of current through wires in pulses. The absence of a pulse represents a 0 and the presence of a pulse represents a 1. Similarly, optical fiber conveys digitally encoded information by turning the light source on and off [3].

Once our speech signal is digitally encoded, it will be transmitted on the same channel as many other simultaneous conversations—a process called *multiplexing.* To understand how digital multiplexing works, imagine 24 digitally encoded conversations flowing into a single T-carrier facility. (Once aggregated, these 24-bit streams make up what in the United States is called a *T1 facility.*) As we have seen, each digital conversation consists of a stream of 8-bit binary numbers. As we have also seen, the standard sampling technique generates 8,000 of these binary numbers (or 64,000 bits) each second [4].

Our 24 separate bit streams flow into a device called a *pulse-code modulator* (PCM), which accepts in turn the first 8 bits from the first conversation, the first 8 bits from the second conversation, and so forth until it has captured the initial numbers of all 24 conversations. Taken together, these 24 numbers are called a *frame.* As soon as the PCM has assembled one of these frames, it begins assembling a second frame made up of the *second* 8-bit number from each of the 24 conversations.

The frames are transmitted in the same order in which they are assembled. In order for the encoded speech of each conversation to arrive at its destination at its original, 64,000-bps sampling rate, the bit rate must be multiplied by 24 (the number of aggregated conversations)—meaning that the frames must move at 1.5 million bps.

Once this high-speed bit stream reaches the distant end of the carrier facility, the process we have described is reversed. The frames are taken apart and the 8-bit numbers are reassembled into their separate conversations. Those conversations are slowed down to the 64,000-bps sampling rate. Finally, at the telephone switch serving the called party, the signal is reconverted to analog form for delivery over the called party's access line.

While this is remarkable enough, digital multiplexing goes much farther. The 24-conversation T1 group can be combined with another T1 to make a 48-circuit group (called a T2) moving at 3.152 million bps, two of those groups can be combined to make a 96-circuit group moving at 6.3 million bps, and so on.

Time-division multiplexing (TDM) is used on digital systems employing wire, cable, microwave, and optical fiber. Regardless of the transmission medium used, the process is the same: the combination of single voice circuits for transmission at a bit rate that equals the number of voice circuits multiplied by the single-circuit bit rate.

Before we leave the subject of MTS technology, we should say a few words about the signaling techniques that control the MTS calling process.

MTS Signaling

Signaling is information transmitted through the telephone network that helps to set up, route, take down, and bill for the call. This information may travel along the same channels as the telephone conversation or may travel on separate facilities devoted exclusively to signaling traffic. As telephone networks become increasingly automated, they rely heavily on rapid, accurate signaling techniques.

The signals with which telephone customers are most familiar are *subscriber loop signals*, which move information between the customer's telephone and the local telephone office. In a typical telephone call, the first subscriber loop signal is the off-hook indication, which (as we saw) is nothing more than the flow of current through the access line that begins when the customer lifts the receiver. This signal alerts the local switch to send the customer a dial tone.

The dial tone, in turn, is a pair of simultaneous tones returned over the access line to the customer. This is the customer's signal to dial a telephone number.

The next subscriber loop signals are sent when the customer dials the telephone number of the called party. These digits are sent over the access line, either as a series of interruptions of the direct current in the line (the method used in rotary telephones) or as combinations of tones that the switch reads as representing numbers (the method used in touch-tone telephones).

Other subscriber loop signals—notably, the busy signal and the ringing signal heard by the calling party—are also transmitted as tones. Unlike the dial tone, these signals are intermittent rather than constant.

Interoffice signals, as the term suggests, convey information between one piece of network equipment and another, rather than between customers and the telephone company. So, for example, when the caller's local switch needs a trunk for long-distance transmission of the call, it will seek a trunk that is emitting an *idle* tone. When the local office *seizes* the trunk for transmission, it turns off the idle signal, thereby alerting the office at the distant end to hold the trunk open for a call. After the trunk is seized, another interoffice signal will transmit the called party's telephone number so that the distant exchange will know which of its subscribers will receive the call. When the called party answers, the distant central office will seize another idle trunk for the return transmission (i.e., the called party's side of the conversation) and will turn off that trunk's idle tone.

Interoffice signaling *may* share the same transmission channel used by the parties' conversation, as subscriber loop signaling does. In this case the interoffice signal may use the same frequency band as the voice signal (this is called *in-band signaling*) or a separate portion of the trunk's bandwidth reserved for signaling (this is called *out-of-band signaling*) [5].

In many applications, however, telephone networks no longer crowd interoffice signaling transmissions onto the facility that also carries the customers' conversations. Instead, they send signals over digital trunks devoted entirely to this purpose. This approach, called *common-channel signaling* (CCS), results in more efficient use of voice facilities. (For example, busy signals are returned to callers without tying up a voice channel that could be used to carry a completed call, and the telephone number of the party making the call can be delivered to the called party.) Efficiency is enhanced by the use of packet switching, which aggregates signaling information associated with many calls and sends it in addressed bursts of data, reducing or eliminating idle time on the CCS trunk [6].

CCS is vitally important to services that use specialized databases to route calls. Advanced intelligent network (AIN) services, for example, permit customers to instruct telephone company computers to route their incoming calls to different locations depending on the time of day. (When a call comes in for the customer, the local switch sends a high-speed query over a CCS link to the AIN database, which returns the number to which the local switch should send the call.) Similarly, AIN permits customers to create virtual private networks and change the configuration of those networks simply by giving appropriate instructions to the AIN computer. Without separate CCS signaling networks, these services would be impracticable.

1.1.2 The Business of MTS

MTS service is still provided in many countries by agencies of the central government. (These agencies are often associated with the post office and are referred to, generically, as postal, telegraph, and telephone (PTT) agencies.) In other countries, MTS is provided by a combination of governmental and private entities, and in still other countries (notably, the United States) the service is provided entirely by private carriers that are regulated by governmental agencies.

Customers may be charged for MTS at a flat rate (i.e., a rate that does not vary with the number of calls placed or the distance those calls traverse) or at metered rates (i.e., rates assessed on a per-call, per-minute, or per-unit-of-distance basis). In practice, MTS rates tend to combine these methods, with calling within some local area available at a fixed rate and long-distance calling provided on a metered basis.

In the United States and a number of other countries, customers may choose one carrier for local service and another for long-distance calls. (Local service is still provided on a monopoly basis almost everywhere.) Customers may also have a choice of providers of telephone equipment, providers of access links to long-distance carriers, and vendors of other MTS-related products and services.

Where MTS service is provided by private regulated carriers, providers may be permitted to set rates that recover their reasonable expenses and a market-based return on capital, or they may be subject to a cap on their rates, but be permitted to increase their profits by reductions in cost. Carriers also may enjoy protection from competition, which is accompanied by limitations on the businesses they may enter. We say much more about these regulatory schemes later in this book.

1.1.3 How International MTS Is Provided

Besides making domestic MTS calls, customers also may dial the telephone numbers of MTS customers in other countries. When this happens, the calling party's local telephone company will recognize the number as designating a foreign destination and will route the call to a switch that is connected to international transmission facilities [7].

In a country that has a single monopoly provider of international service, the international call will be routed to that provider's *gateway* switch and from that switch to an international cable or satellite circuit. In a country that has more than one international carrier, the caller's local switch will execute a procedure for choosing among those carriers. In the United States, for example, the call typically will go to an interexchange carrier to which the caller already has subscribed, and that carrier will route the call to a gateway switch. (If the customer's interexchange carrier does not provide international service, then that carrier will hand the call to a carrier that does provide such service.)

Once the international carrier has the call, its switch will read the initial digits of the telephone number to identify the destination country. If the international carrier has a *correspondent relationship* with a carrier in the destination country, it will send the call over an international facility to the foreign carrier for delivery to the called party [8].

If the sending and receiving countries share a common border, the call may travel over a facility similar to one that connects switches within a telephone company (either a metal cable, a terrestrial radio link, or a fiber-optic cable). If the two countries are widely separated, and particularly if they are separated by an ocean, the call will travel by undersea cable or by satellite. We should say a few words about each of these transmission technologies.

Undersea Cable Technology

Undersea cables have proved to be a reliable means of carrying transoceanic MTS conversations. They are also remarkable feats of engineering and construction.

Undersea cables are laid by specially designed cable ships. Along coastlines and in shallower regions of the sea, cables are laid in special trenches dug in the ocean floor. In deep ocean, the cables are simply trailed behind the ship and allowed to sink to the bottom. Because repair of deep-ocean cables is slow and expensive, both the cables themselves and the repeaters that regenerate the signal every several miles must be of exceptionally rugged and reliable design.

Although undersea telegraph cables date from the mid-nineteenth century, transoceanic cables suitable for telephone traffic did not enter service until the 1950s. The first generation of undersea telephone cables used coaxial technology and vacuum-tube repeaters. The second generation was also based on coaxial cable, but replaced the vacuum-tube repeaters with transistorized equipment. Improvements in transmission technology increased the capacity of these coaxial facilities from 72 two-way circuits (on the TAT-1 and TAT-2 circuits laid in the late 1950s) to 10,000 two-way voice circuits on the cables laid in the late 1970s and early 1980s.

A much greater breakthrough occurred in 1988, when AT&T and 28 European and North American partners laid the TAT-8 fiber-optic cable. This transatlantic cable accommodated 40,000 simultaneous voice conversations—four times the capacity of the most efficient transoceanic coaxial cable—and required fewer repeaters to regenerate the signal. It also had the capacity to carry video as well as voice and data signals [9].

Since TAT-8 was laid, various interests have placed a number of high-capacity fiber-optic cables across the Atlantic and Pacific.

In the Atlantic, AT&T and other investors have laid the TAT-9, TAT-10, and TAT-11 cables. TAT-9 can carry twice as much traffic as TAT-8 and lands in Spain as well as France, the U.K., and the U.S. TAT-10 connects the U.S. with Germany and the Netherlands, and TAT-11 connects the U.S. with France and the U.K. Taken together, the TAT cables, which are available as backups to each other in the event of service interruption, offer a high degree of reliability and routing diversity for transatlantic communications.

A group of purely private investors (i.e., not including any PTTs) has laid a third fiber-optic cable across the Atlantic. Branches of this cable, known as PTAT, run from the United States to Ireland, the United Kingdom, and Bermuda. The system is owned by Cable & Wireless in the United Kingdom and Sprint in the United States. PTAT, with a capacity of 100,000 two-way voice circuits, carries MTS along with digital switched data services, private lines, and virtual private (software-defined) networks.

The first of the transpacific cables, TPC-3, was financed by a large consortium of public and private carriers and runs from the U.S. mainland to Hawaii with branches extending to Guam and Japan. When this system entered service in 1989, it was possible for the first time to send a message over a fiber-optic facility from Europe to Japan. Like TAT-8 in the Atlantic, TPC-3 can carry 40,000 two-way voice conversations.

Since TPC-3 entered service, a number of additional transpacific cables have been laid or have been scheduled for construction. Notably, TPC-4, owned by a consortium of 31 carriers, runs from Japan to British Columbia and California, and TPC-5 will link the U.S. to Japan by way of Hawaii and Guam. Eventually, TPC-5 will also be linked to the planned Asia Pacific Cable Network (APCN), which will connect Brunei, Guam, Japan, Indonesia, Korea, Malaysia, the Philippines, Singapore, Taiwan, and Thailand.

Finally, the most ambitious fiber-optic cable project of all is the Fiberoptic Link Around the Globe (FLAG), which will connect Europe, the Middle East, and Asia. FLAG is owned by NYNEX and five partners and is scheduled to enter service in September of 1997.

With the successful implementation of fiber-optic technology, undersea cables are a credible competitor to satellite circuits for international transmission of all types of telecommunications services.

Communications Satellite Technology

Most of the communications satellites in use orbit the Earth at an altitude of 22,300 miles. At this altitude their orbits are synchronized with the rotation of the Earth about its axis, so that these satellites always occupy the same position in relation to the Earth's surface [10].

Communications satellites carry dozens of *transponders*—radio devices that receive, amplify, and retransmit microwave radio signals received from terrestrial radio stations. The signal sent from Earth to the satellite, called an *uplink*, is a narrow, focused signal. The signal retransmitted to Earth by the satellite, called a *downlink*, is on a different frequency from the uplink and may reach Earth stations spread over a wide area.

The first generation of communications satellites used the same radio frequencies as terrestrial microwave systems—a group of frequencies called the

C band. A newer generation of satellites uses a group of higher frequencies called the *Ku band*.

All satellites using the same frequency band must be separated at a distance that will keep their signals from encountering each other, either in space or at the antennas of their respective Earth stations. For C-band satellites, this requires a separation of 4 to 5 degrees (or about 2,000 miles) between satellites. For Ku-band satellites, which transmit on narrower high-frequency signals, spacing of 2 degrees is sufficient. (We say more later in this book about the allocation of scarce orbital slots among competing claimants.)

The principal limitation of satellites as a transmission medium—especially for MTS service—is delay. It takes about 270 ms, or something over one quarter of a second, for a signal to travel 22,300 miles into space, change frequencies for retransmission, and return 22,300 miles to the receiving Earth station. While this delay is unimportant for one-way transmissions, it is highly distracting to people trying to converse on an MTS circuit, who often find themselves "talking over" one another. For this reason, satellites are best adapted to video and data applications.

This does not mean, however, that satellites are not an important source of capacity for MTS transmissions. Some carriers find that the delay problem can be reduced by carrying one side of the MTS conversation by cable and the other side by satellite, and many carriers look to satellites as a valuable backup to undersea cables [11].

As we shall discuss at greater length later, transponder capacity for international transmissions is available from a number of satellite systems. The principal source is INTELSAT, which owns and operates 23 satellites that serve over 300 users. Satellite capacity is also available, however, from so-called separate satellite systems [12].

Completing the International Call

Once our international call has reached the gateway switch in its country of origin and once the switch has selected an undersea cable or satellite circuit to carry the call, all that remains is to send the call to the destination country for delivery to the called party. But nothing in telecommunications, of course, is quite *that* simple.

First, the originating carrier must monitor the call for billing purposes so that it can charge the caller and make accurate payments to the receiving carrier under the correspondent agreement.

Second, the gateway switch may have to modify the voice signal to make its audio level compatible with that of the receiving country's network so that the customers will not be distracted by echo and feedback during their conversation.

Third, the gateway may have to convert the signaling system of the originating carrier to one suitable for international transmission. (As we have seen, the digits of the called party's telephone number, busy and ringing signals, and interoffice signals of various kinds must move through the network in order to set up, route, and take down an MTS call.) In spite of the trend toward standardization of signaling techniques around the world, gateways will continue to make these translations for some years to come.

1.2 WIDE AREA TELECOMMUNICATIONS SERVICE

Wide Area Telecommunications Service (WATS) is a voice telephone service invented by AT&T and now offered by a number of carriers in North America, Europe, and Japan. One version of the service, usually called Outward WATS, permits subscribers to place outbound calls to a variety of distant locations at a bulk rate. Another version, known variously as Inward WATS, 800 service, or, on the European continent, Green Number service, permits callers from prescribed areas to reach a business number without paying for the call. (These latter services, of course, have become an enormously important marketing tool for businesses of all kinds.)

While Outward WATS is primarily a billing arrangement, Inward WATS is the most successful example of a telephone service based on intelligent network technology. When a caller dials an 800 or other Outward WATS number in the United States, for example, the local switch sends a query over a CCS signaling network to a database that locates the telephone number to which the customer wants its Inward WATS calls sent. The database sends that number back to the local switch, which routes the call accordingly.

Some carriers now offer international Inward WATS calling, under which a customer pays for inbound calls to his or her number from a designated foreign country or countries.

1.3 PRIVATE LINES

So far, our description of telecommunications services has focused on *circuit-switched* services (i.e., services provided over temporary connections established by carrier-provided switches). Business customers, however, often purchase *dedicated* transmission facilities connecting locations between which they have heavy communications needs. These facilities, called *private lines*, are reserved for one customer's needs at all times, and the charges for these lines do not vary with the amount of traffic they carry.

For customers with significant telecommunications traffic among a number of locations, private lines offer both economy and simplicity: economy because a flat charge for a dedicated link may be less than the metered charges the customer would incur if it made the same number of calls on an MTS basis; simplicity because locations on a private network can be reached with numbers of 7 digits or fewer, rather than the 10 or more digits needed for domestic or international direct distance dialing.

In some countries (notably the United States, the United Kingdom, and Japan), private lines are available from competing providers; in others, they are a monopoly service of the PTTs. Acquiring international private lines can be cumbersome if the customer tries to deal separately with the carriers at each end of the circuit. Ordinarily, the customer must purchase a *half-circuit* from each carrier, with the half-circuits meeting at the theoretical midpoint of an undersea cable or at the satellite through which a satellite circuit will pass. Then the customer will pay each carrier separately for that carrier's portion of the service. Where available, it is much more convenient to purchase the private line from a single carrier, which will make its own arrangements with the corresponding carrier and present the customer with a single bill for the whole circuit.

Private lines may be digital or analog, broadband or narrowband, depending on the customer's needs.

On the analog side, the basic private line is a single voice-grade facility that reliably utilizes the full frequency range of a normal telephone circuit. (These lines are useful for voice and for data moving at rates of up to 9,600 bps.) Customers can also order less expensive analog lines that are suitable only for voice and lower speed data transmissions, and can purchase lower-than-voice-band facilities that only support teletype and other extremely low-speed data applications.

Analog private-line facilities also are available in bandwidths greater than that of an ordinary voice line. The standard is the 48-kHz facility, which can be used for high-speed data or can be multiplexed by the user (on a frequency-division multiplexing (FDM) basis, as discussed earlier) into a number of voice channels.

On the digital side, private-line facilities in various bandwidths are available within and among most developed countries. Among other differences between these and the analog private lines, these facilities include a digital link that extends all the way to the customers' premises, replacing the analog access line. Because of this, it is not necessary for the user to place a modem on the access line to convert the digital signal from the sending device into a stream of analog tones.

1.4 SOFTWARE-DEFINED NETWORKS

Software-defined networks (SDN) were introduced by AT&T in the late 1980s. They offer all of the advantages of dedicated private-line networks, but have greater economy and flexibility.

The key to SDN service is common-channel signaling. SDN calls are carried over the SDN provider's switched network, but call routing is controlled by an SDN database rather than the carrier's MTS switches. Using this approach, the customer can dial abbreviated numbers to reach other locations on the network, just as though the network used dedicated lines. The database simply translates the abbreviated numbers into the ordinary MTS numbers of the various network destinations.

The SDN is not only as convenient as a network made up of private lines; it may be more economical and flexible. This is so because locations that lack the traffic to justify a dedicated line can be connected to a virtual network at reasonable cost and because the network configuration can be altered by a simple programming change rather than removal and installation of physical links.

As fiber-optic cables and signaling system 7 (SS7) signaling [13] are deployed internationally, true international SDNs will become available to connect locations in North America, Western Europe, and Japan.

1.5 PACKET-SWITCHED SERVICES

We spoke of packet switching earlier when we discussed the common-channel signaling networks used by telephone companies. As we pointed out in that connection, packet switching achieves enormous efficiencies by aggregating signals from many calls and sending them in separately addressed bursts, or packets, over a single facility.

Packet switching is useful, not only in telephone signaling, but in all types of data transmissions [14]. To understand why this is so, we should explain this technology in more detail.

First, imagine yourself at your computer, making a query to a remote database. Your computer might be connected to the database by a temporary voice-grade MTS circuit established by your telephone company, or it might be connected to the database over a private line. Either way, the line will sit idle for most of the time you are using it. It will sit idle while you compose your request, while the host computer processes your request, while you ponder the information the computer sends you, and while you compose another request. The only time the line does any work is when it transmits your request and transmits the response.

Now, suppose instead that your request goes to a switch that aggregates it with many other, roughly simultaneous requests and sends all of those requests in a continuous series of addressed bursts—or packets—of bits. Now only the line from your computer to the packet switch experiences idle time; for most of the transmission path, the line is moving data most or all of the time.

Packet switching, by aggregating many data messages in this way, reduces the cost of data transmission dramatically. Instead of paying for a complete, end-to-end circuit while you talk to the database, you pay only for transport of a few bursts of data over a facility whose cost is shared with many other users. This is a packet data service.

A complete packet service requires a device called a packet assembler-disassembler (PAD) at each end of the line to combine outgoing packets and disaggregate incoming packets, a packet switch to address and send the aggregated packets, and a high-speed data facility to carry the packetized messages.

Some packet-switching users (notably, the telephone companies) can justify the expense of operating all of this equipment themselves; but for those whose needs are more modest, a number of companies now provide packet data network services on a common-carrier basis. Some of these networks (including those provided by several PTTs) have international capability.

Transmission of data over packet networks may involve some complex processing problems, which are discussed at greater length in Chapter 2. Notably, some users may operate computer networks based on proprietary protocols that cannot "talk" to computer systems based on other protocols. To overcome this problem, some packet data service providers offer a service called *protocol conversion*, which automatically translates among different networking protocols. (As we discuss later, international standards—notably, CCITT Recommendations X.25 and X.75—have been developed to ease or eliminate internetwork translation.)

1.6 FRAME RELAY AND CELL RELAY

Packet-switching systems, while far more efficient for data transmission than the circuit-switching methods of the voice telephone system, suffer from a crucial limitation in transmission speed. Specifically, at various points (called *nodes*) in a packet-switched network, the system scrutinizes each packet for errors, corrects the errors or directs that the packets be retransmitted, and reassembles the packet before passing it on to the next node. The delay caused by this process limits packet networks to speeds of about 64 Kbps, or 64,000 bps.

One modern variant of packet switching, called *frame relay*, addresses this problem by sending data without pausing to check for errors, relying on the customer's terminal equipment to find errors and request retransmission if necessary. Frame relay systems transmit data at speeds of 1 to 2 Mbps. This technology is best suited to pure data applications, including interconnection of local-area networks (LAN).

Another variant of packet switching, called *cell relay*, transmits packets (called *cells*) of fixed length that are easier to switch. Cell relay systems are capable of speeds of hundreds of megabits per second. The cell relay approach, like frame relay, is well-adapted to pure data transmission [15], but because of its greater speed it also offers the potential for switching voice, data, and video through the same switch [16].

1.7 TELEX SERVICE

Telex service was first offered by the Western Union Company in 1958. It requires specialized equipment at each end of the transmission and permits direct communication only among subscribers to the service.

A telex message is typed on a keyboard. The characters of the message are then converted into coded electronic impulses and a teleprinter at the distant end types a printed version of the message. Any telex subscriber can reach any other subscriber on a direct-dial basis. When the recipient of the message is not a telex subscriber, the message can be transmitted to the nearest office of the telex carrier and hand-delivered from there as a telegram.

Telex service is rapidly being replaced by facsimile and electronic mail, but it is still an important means of communication in Eastern Europe and developing nations.

1.8 FACSIMILE SERVICE

Facsimile (or simply *fax*) service, like telex service, requires specialized equipment to send and receive messages, but does not require users to subscribe to transmission services provided by any particular carrier or equipment provider.

Users initiate fax transmissions by dialing the telephone number of a receiving fax machine, then feeding a document into the sending machine [17]. The fax machine (a combination telephone, scanner, and modem) scans the document optically, converts the light and dark elements of the document into a stream of digital bits, and converts those bits into analog tones suitable for transmission over the telephone access line. From this point the telephone network carries the fax transmission exactly as it would an ordinary MTS call. When the transmission is delivered to the receiving machine, it is printed as a reasonable replica of the original document. Customers pay for fax transmissions—whether domestic or international—on the same basis as they would pay for an ordinary MTS call of comparable distance and duration [18].

As we describe more fully in Chapter 2, international fax service has benefited from the development of data protocols that ensure compatibility between sending and receiving networks and fax equipment.

1.9 ELECTRONIC MAIL

We typically send electronic mail (e-mail) by typing a message on a computer keyboard and sending that message over a private or public network to another computer user, who then can read the message on the screen of his or her computer [19].

Many e-mail networks are private; that is, they are offered within a LAN in a single office or building, or within a wide-area network connecting a number of institutional or corporate locations. But a number of carriers—including AT&T, MCI, Sprint, and British Telecom—are offering domestic and international e-mail service on a common-carrier basis. These public services are growing rapidly, and e-mail is at the threshold of becoming ubiquitous in the business world. (Already, many letterheads and business cards display e-mail addresses along with street addresses and telephone numbers.)

As we discuss in more detail in Chapter 2, the growth of international e-mail is heavily dependent on the availability of methods of translating messages between one e-mail system and another. The principal such *protocol*—Recommendation X.400—was established by the International Telecommunications Union (ITU) and has achieved wide acceptance among both private and public e-mail systems.

1.10 INTEGRATED SERVICES DIGITAL NETWORK

If the fondest hopes of its advocates are realized, the integrated services digital network (ISDN) will bring the benefits of digital transmission and common-channel signaling to the premises of small business and residential customers without the need to replace their present access lines with coaxial cable or optical fiber, and will permit the worldwide interconnection of private and public telecommunications networks over facilities using a single standard.

As we explain in more detail in Chapter 2, ISDN began not as a service, but as an international technical standard. The basic ISDN standard calls for delivery to the customer's premises of three digital channels: two 64-Kbps circuit-switched lines for the digital transport of voice and high-speed data and a 16-Kbps packet-switched line for signaling and lower speed data transmissions. Both voice and data would move between the customer and the central office over these channels using the existing twisted-pair copper wire access line. Multiplexing and demultiplexing would be handled in the customer's terminal equipment.

ISDN may make a number of services, including videoconferencing and high-speed data exchange, available to residential and small-business customers. The only limitations on ISDN's prospects are the willingness of customers to pay for these new services and the possibility that newer technologies will overtake ISDN before it can be widely deployed.

1.11 CONCLUSION

In the chapters that follow, we talk about the rules by which the services we have described (and some other services we introduce later) are provided within and among carriers and nations. As we work through that process, we will also build on the descriptions given in this chapter to provide a more detailed picture of the evolving global telecommunications network.

Notes

[1] Digital transmission facilities are designed to sample voice frequencies up to 4,000 Hz. (In fact, the human voice includes overtones higher than this, but those higher frequencies can be filtered out without noticeable loss of quality.) Telephone systems generally have selected 8,000 measurements per second as their sampling rate, based on a theorem (the Nyquist theorem) that states that any given sampling rate will capture all the information contained in frequencies below one-half the sampling rate. Put another way, it is necessary to sample analog information at twice its highest frequency component.

[2] Carrier systems are the facilities used to transmit aggregated traffic between telephone switches on a multiplexed basis. These facilities are distinguished from the access line connecting your telephone to the central office, which is a simple pair of copper wires carrying direct current (not so different, really, from Bell's first telephone), and from local cables that carry a large number of these twisted pairs for short distances without multiplexing their signals.

[3] Transmission systems that use radio frequency (RF) waves to carry digital signals, such as microwave radio, also represent each bit as one of two states of the signal. But in those systems the two states are not achieved as they are in T1 and optical-fiber systems by simply turning the signal on and off. Instead, the radio wave is altered, either in amplitude, frequency, or phase, between two states—one state representing a 1, the other representing a 0.

[4] Many telephone systems still use analog rather than digital carrier facilities in part or all of their networks. Where analog facilities are used for interoffice transmission, voice signals may be multiplexed by using an RF signal to carry the voice waveform and sending several such signals at once at different frequencies. (RF waves can share the same channel so long as they are transmitted at different frequencies.) This analog multiplexing technique is called *frequency-division multiplexing* (FDM). Digital multiplexing is called *time-division multiplexing* (TDM).

[5] A transmission facility's bandwidth is its capacity to carry information. For an analog facility, bandwidth is a function of the range of frequencies the facility can carry. For a digital facility, bandwidth is a function of the rate at which the facility can transmit data.

[6] We say more about packet switching in Section 1.5.

[7] We describe the international dialing plan in Chapter 2.

[8] A correspondent relationship typically involves an agreement to share revenues from calls between the correspondents' networks and the establishment of a partnership or other joint interest in the international facility used to carry those calls.

[9] TAT-8 extends east from New Jersey and splits into branches that land in France and the United Kingdom. Most of the capacity of TAT-8 is used for ordinary telephone traffic carried by the cable's owners, but some of the capacity is used by large customers for private-line services. The TAT-8 system consists of two active optical-fiber pairs and one backup pair.

[10] Low-Earth-orbit (LEO) satellites are also entering service for use in conjunction with new types of mobile telecommunications services.

[11] While undersea cables are highly reliable, outages on these facilities take much longer to repair than interruptions in service on terrestrial facilities.

[12] For a thorough discussion of INTELSAT, see Chapter 4.

[13] See Chapter 2.

[14] A form of packet switching called *asynchronous transfer mode* (ATM) will enable networks to carry both voice and data on a packet-switched basis.

[15] As a method of pure data transmission, cell relay is the basis for switched multimegabit data service (SMDS), a public data service used primarily to connect LANs with each other.

[16] Used for simultaneous voice, data, and video switching, cell relay is the basis for ATM technology.

[17] Many computers now contain fax modems that transmit documents stored in the computer's memory.

[18] Some carriers are offering special services—such as store-and-forward capability—for use in connection with fax machines. These services may not be available through an ordinary access line, since they require a dedicated link to the carrier providing the service.

[19] While this description covers e-mail as most of us know it, some e-mail services offer voice mail and imaging as well as text.

International Telecommunications Standards

2

When we transmit information electronically from one place to another, we have at least two expectations for it: it will arrive and it will be understood. In order for these expectations to be fulfilled, the facilities and systems at each stage of the transmission must be technically compatible. All equipment along the transmission path must be linked by adequate physical connections, and the transmission's format and content must be appropriate for the systems and equipment over which it will travel.

Until recent decades, technical compatibility along most transmission paths could be taken for granted. Within countries, public telecommunications were provided by PTTs or private monopolies that maintained uniform equipment and transmission standards nationwide. Between countries, telecommunications consisted largely of analog voice conversations, telex, and other transmissions that offered few compatibility problems and could be harmonized readily at the gateways through which international traffic moved. And networking of computers, either on private systems or through the public telephone network, barely existed.

The present environment, of course, is entirely different. As nations privatize and liberalize their telecommunications sectors, public networks must increasingly interconnect with customer-provided equipment, private networks, value-added services, and competing carriers of all kinds. International communications must accommodate digital voice transmission, out-of-band signaling, high-speed data communications, and other technologies that require a high degree of intersystem coordination [1]. And computer equipment and networks are increasingly linked with other equipment and networks that may use different, proprietary communication protocols.

One method of achieving compatibility among this welter of equipment and systems is to connect them through interfaces that translate among them. So, for example, international gateways convert incoming and outgoing signals to accommodate the standards of different national networks, and computers using different communication protocols can be connected through *protocol conversion* services that translate the sending computer's protocol into the receiving computer's protocol. But the interface approach—even where technically feasible—is a second-best solution that adds costs and reduces flexibility. The better solution is for all equipment manufacturers and system operators to make their equipment and networks compatible—that is, to *standardize* those features and specifications that affect interconnection and interoperability (see Figure 2.1).

Where standardization is achieved, users and providers of telecommunications equipment and services benefit in at least two ways. First, standardization can reduce the cost of equipment and services and increase the options available to consumers. This is the benefit offered, for example, by standardized computer networking protocols such as the X.25 standard. Where computers and networks support such a standard, the products of many different vendors may be networked efficiently: in the absence of such a standard, users who wish to connect their equipment and networks with those supplied by different vendors may have to replace their equipment or assume the added cost of a protocol conversion service.

Second, standardization can prevent firms with market power from using that power to impede competition. This is the purpose, for example, of the FCC's Part 68 equipment registration program, the European Union's (EU) Open Network Provision (ONP) regulations and the FCC's *Computer III* regulations. All of these programs seek to ensure that vendors of equipment, value-added services, and other competitive offerings can connect their equipment and services to the public network on fair and efficient terms that are equivalent to the arrangements the dominant carriers provide for themselves.

2.1 COMMON PROBLEMS WITH THE PROCESS

However desirable, standardization is not a tidy process and its aims are not always perfectly achieved. Ideally, a single, omniscient body would anticipate all of the compatibility problems posed by emerging communications technologies and would solve those problems with timely, efficient standards that benefit all interested parties equally. In reality, of course, standards are not the result of a single, defined process. Some standards are imposed by government regulation, others are adopted voluntarily through the participation of interested parties in national and international standards-setting organizations, and still others begin as proprietary standards of individual manufacturers or service providers and achieve dominance because of the success of their proponents in the marketplace. (Standards that emerge in any of these three ways are known, respectively, as de jure, voluntary, and de facto standards [2].) International voluntary standards-making bodies often come late to these processes and are impeded by the presence of conflicting de facto and de jure standards within their own entrenched constituencies [3]. In

Type of Standard by Goals

Standardization mechanism	Control	Product/quality	Process/interoperability
De Facto	Warner-amex Database-privacy standards	VCR standards	Language customs Bills of lading Computer interface standards
Regulatory	Auto safety regulations Fuel economy standards	NSA encryption standards Department of Agriculture Product classification standards	Open network architecture standards ETSI standards for European telecommunication standards
Voluntary consensus process	Standards for medical devices Pressure vessel standards Petroleum standards	Refrigerator standards	Map-top protocols for OSI/standards Standards evolving legislation Electronic data interexchange standards

Figure 2.1 As this chart from a U.S. government publication shows, telecommunication and data communication standards tend to be classified as process/interoperability standards and may be de facto, regulatory, or voluntary. (*Source:* U.S. Congress, Office of Technology Assessment.)

these cases, the international organizations may be unable to fashion a single standard or may have to accept a solution that is not technically optimal.

While an exhaustive discussion of the problems of standards making is beyond the scope of this chapter, some examples will give a flavor of the principal obstacles that separate the real from the ideal.

2.1.1 Anticompetitive Abuse of the Process

As we noted earlier, one purpose of the standards process is to promote openness and discourage abuse of market power; but standards may become a tool of monopolists rather than a means of restraining them. This can happen in a number of ways, but usually where a monopolist preempts the voluntary and de jure processes and imposes an anticompetitive de facto standard, or where the processes of a standards-making body are misused for anticompetitive purposes.

An example of anticompetitive de facto standards making occurred after the FCC ruled that telephone subscribers in the United States would be permitted to attach their own equipment to AT&T's monopoly network. AT&T and its Bell operating companies (BOC) insisted that any non-Bell equipment had to meet rigorous standards of physical and electronic compatibility with the Bell network. Nonconforming equipment, according to Bell, could destroy the switched telephone network [4].

AT&T's competitors conceded that interconnection standards might be useful, and they argued that those standards should take the form of minimum performance criteria enforced by an FCC registration program. Under such a program, manufacturers could obtain the right to attach their equipment to the network by certifying the equipment's compliance with the reasonable, published standards.

In the interim before the FCC acted on this suggestion, however, AT&T announced its own interconnection standard. Instead of proposing a set of design criteria, AT&T adopted an "interface" approach—a coupling device, sold only by AT&T, that supposedly ensured that no harmful transmissions would pass from non-Bell equipment to the public network. As a number of courts later found, AT&T's so-called protective coupling arrangement (PCA) was an effort to preempt the standards-making process and inhibit competition [5]. Because the PCAs were expensive and not always available, they increased the cost and delayed the sale of competing equipment.

Eventually, of course, the FCC did enact a registration program and the PCA approach was abandoned—but not before AT&T had incurred antitrust liability to a wide variety of equipment manufacturers. The PCA experience, along with AT&T's efforts to make interconnection with its network difficult for competing long-distance telephone companies, convinced the FCC that the terms and conditions of interconnection to the U.S. public network could not be imposed on a de facto basis by the dominant carriers; they had to be specified by a combination of voluntary and de jure standards that applied equally to the established carriers and their competitors. Accordingly, the FCC, while not itself writing detailed technical specifications for interconnection, has enacted elaborate systems of rules requiring equal access for long-distance companies and enhanced (value-added) service providers, and requiring full disclosure of any changes to the network that will affect interconnection and interoperability with equipment and services provided by outside vendors.

Misuse of standards-making bodies occurs when the organization itself becomes a conspiracy of its members against nonmembers or when some members "capture" the organization and use it to the detriment of other members. Some of the best examples of this phenomenon have occurred not in the telecommunications industry, but in other industries that have come under the scrutiny of U.S. antitrust laws.

A conspiracy of members against nonmembers was found, for example, in the case of the American Society of Mechanical Engineers (ASME), Inc., v. Hydrolevel Corp. [6]. ASME is a publisher of over 400 codes and standards affecting a variety of industries in North America. One of ASME's members—McDonnell & Miller (M&M), a manufacturer of cutoff valves for heating boilers—became concerned when a competitor introduced a new type of cutoff valve and sold it to one of M&M's customers. M&M decided to use its membership in ASME to discredit the new product and arranged with the chairman of one of ASME's standards committees to obtain a written disapproval of the new product's safety. M&M then made extensive use of the written opinion in its marketing efforts. A jury found that the written opinion was not issued in good faith and that the conspiracy between M&M and various members of ASME's standards-making committees was an anticompetitive abuse of ASME's processes [7].

Even when a standards-making body includes all interested parties and makes an effort to ensure procedural fairness, the processes of the organization may be used by some members to the detriment of others. In another case that resulted in an antitrust decision by the U.S. Supreme Court, a manufacturer of plastic electrical conduit alleged that one of its competitors—a maker of metal conduit—had subverted the processes of the National Fire Protection Association (NFPA), a private organization that promulgates an influential set of standards called the National Electrical Code. According to the plaintiff, the Allied Tube & Conduit Corporation "packed" the NFPA's annual meeting with attendees from the steel industry who had not previously been members of the NFPA and otherwise "subverted the consensus standard-making process of the NFPA" [8]. The result of Allied's campaign was the defeat of a resolution approving the plaintiff's plastic conduit as safe for use under the National Electrical Code. The jury found that that while Allied's behavior at the annual meeting was technically in compliance with NFPA procedures, its aggressive orchestration of the steel interests' participation in the meeting overwhelmed the plastic conduit manufacturers and subverted the process. As the Supreme Court noted in affirming the decision for plaintiff of the Court of Appeals, the propriety of the conduct of a member of a standards-setting organization "is not established, without more, by [the member's] literal compliance with the rules of the Association, for the hope of procompetitive benefits depends upon the existence of safeguards sufficient to prevent the standard-setting process from being biased by members with economic interests in restraining competition" [9].

While we have not told similar tales of abuse of the principal standards-making bodies that affect the telecommunications industry, this does not mean that those organizations never act against nonmembers or are free of competitive rivalries within their membership. Some commentators, for example, have suggested that the ITU [10] displays the bias of its members (i.e., the public networks of the member nations) by designing an ISDN system that "does not contemplate an interface at which a subscriber, or an independent service provider, can obtain access to the system without employing the telephone company's local loop" [11]. According to these critics, the ITU's approach to ISDN serves the PTTs' interest in encouraging user dependence on the local loop and works against the desires of independent suppliers and users to have access to the network at as many points as possible [12].

2.1.2 Entrenched Proprietary Standards

As we noted earlier, the international standards-making process sometimes addresses a particular compatibility problem only after a de facto standard has already emerged. If influential telecommunications networks and suppliers have made a significant investment in the existing standard, it may be pointless for international organizations to adopt a conflicting standard, even when a different standard will be more efficient.

Perhaps the best-known example of this problem had its origin in the 1950s, when AT&T began to develop its TDM method for digital transmissions over the telephone network [13]. AT&T opted for the technology that came to be called the *T1 system*, and that technology was ordered in quantity by the BOCs beginning in 1962.

The AT&T approach was well entrenched in North America by the time the responsible ITU committee—the International Telegraph and Telephone Consultative Committee (CCITT)—began to look in earnest at developing its own standard [14]. By that time, the CCITT was under some protectionist pressure from its European membership, which knew that North American manufacturers offered the only digital transmission equipment then available. If European manufacturers were to have a reasonable chance of participating in the conversion of European systems to digital transmission technology, a little incompatibility between the North American and European standards would not hurt.

For this reason, and for engineering reasons as well, the CCITT standard differs from the North American standard in two important respects. First, the two standards use different methods of converting analog voice signals to digital signals and back again. As a result, if an analog signal is converted to a digital signal under one standard and then is reconverted to an analog signal under the other standard, a loss of signal quality occurs. Second, the CCITT and North American digital multiplexing techniques use different multiplexing hierarchies, and these differences become significant at higher digital transmission rates. Both of these differences require international users to ensure that appropriate conversions are performed on international circuits where the differences affect transmission performance.

The story of the North American and CCITT digital transmission standards offers important lessons in the difficulties of overcoming not only existing de facto standards, but the power of competing national and commercial interests. Fortunately, the next generation of digital transmission systems will be based on fiber systems and will benefit from the Synchronous Optical Network (SONET) standards that are designed to achieve interoperability among North American and European systems.

2.1.3 Differences Among Domestic Regulatory Regimes

The development and implementation of voluntary international standards are sometimes hindered by differences among the regulatory regimes of the countries that will use the standards. Perhaps the best-known example of this constraint occurred in the development of ISDN.

The CCITT envisioned ISDN as an end-to-end digital network that users could access through a standard set of interfaces. Part of the task of ISDN standards making, therefore, was to identify those points of interconnection between users and the network—called *reference points*—for which ISDN access standards would be specified. The CCITT established these reference points at the customer side of equipment located at the customer's premises and furnished by the ISDN service provider. (The equipment, in turn, is referred to as the *network interface.*)

While the CCITT was finalizing these specifications, the FCC was establishing a regulatory regime in the United States that prohibited carriers from supplying customer premises equipment to end users as part of an end-to-end service. This made the CCITT-defined interfaces unusable in the United States and required U.S. standard makers to develop a set of specifications for the design of customer-premises equipment that could be used to establish a noncarrier ISDN interface. The standards for this so-called U interface were developed by Committee T1 of the Exchange Carrier Standards Association over a period of about four years and were completed in time for inclusion in the "Blue Book" issued by the 1988 Plenary Session of CCITT. While the standards drafted by Committee T1 permit the deployment of CCITT-compatible ISDN networks in the United States, the task of adapting ISDN to the FCC's rules was a substantial, costly detour [15].

2.1.4 Standards Overtaken by Events

It is often said that the fairest, most efficient way to define a standard for a networked service is to do so a priori—that is, with the participation of all interested parties and before the service has been brought to market under a vendor's proprietary, de facto standard [16]. One problem with this approach in dealing with industries undergoing rapid technological change is that a lengthy, deliberate standards-making process may become irrelevant before it is complete.

Some believe, for example, that this will prove to be the fate of ISDN. While the CCITT has been defining a digital communications system based on the public network and a set of standard interfaces, the computer industry has developed its own LAN services and standards that offer low-cost, high-speed data communication services for business customers. Neither these services nor the growing market for LAN-to-LAN and other internetworking arrangements is developing as an ISDN-based service, raising the possibility that ISDN's future must be found in the residential market.

The rest of this chapter looks at the institutional and procedural background of the standards-making process and then looks at some standards that have particular importance for international services.

2.2 INSTITUTIONS AND PROCESSES

The story of formal, multinational standards making for telecommunications service is primarily the story of the ITU and its consultative committees. While we also have occasion to discuss other organizations, therefore, the ITU will take the lion's share of our discussion.

2.2.1 The History of the ITU

The ITU began as the International Telegraph Union, which was founded at a conference of 20 European countries in 1865. The stated purpose of the organization was to universalize telegraph service among nations. While its membership at first included only European states, countries from other continents were represented within a short time after the organization's founding.

Until the First World War, the ITU had only a modest permanent organization. Between conferences of the membership, a small secretariat in Berne, Switzerland, collected telegraph statistics, published a journal, and provided a place for ad hoc meetings of technical experts.

In 1925, the ITU took the important step of establishing the first of its consultative committees—the International Telegraph Consultative Committee and the International Telephone Consultative Committee, which together came to be known as the CCITT. With these organizations, the ITU could devote continuous attention to the technical and operational problems of international telephone and telegraph service.

For much of its history, the ITU was conservative in its acceptance of nontelegraph technologies. The telephone, for example, was invented in the 1870s, yet was largely ignored by the ITU until 1903. (The view of most delegates, when the matter first came up at the Berlin Conference in 1885, was that international telephone service was in its infancy and adequately covered by bilateral agreements.) Even at the 1903 conference, the ITU disposed of telephones by enacting a short supplement to the existing telegraph regulations. A separate set of telephone regulations was not adopted until the Madrid Conference of 1932.

Radio also came late to the ITU's agenda, although not for the same reasons. A separate organization—the International Radiotelegraph Union—had dealt with radio regulation until 1932, when that organization merged with the ITU. In that year, the ITU acquired the International Radio Consultative Committee (CCIR) and became the International Telecommunications Union.

Radio confronted the ITU with a new problem—the management of frequencies on the electromagnetic spectrum. Radio waves, unlike electricity flowing in a telephone or telegraph wire, do not follow a narrow path from sender to receiver. They are propagated into the atmosphere, traveling great distances and crossing international boundaries with ease. If two such radio signals are propagating in the same place at the same or adjacent frequencies, both signals may experience interference.

Before its merger with the ITU, the International Radiotelegraph Union had begun the work of assigning frequencies to particular services: a group of frequencies had been reserved for coast stations, another for ships, yet another for broadcasting, and so forth. The ITU continued this work, which grew more troublesome as the airwaves grew more crowded.

One problem the ITU confronted for much of its history was the attitude of North America toward its work. Telegraph and telephone service in the United States and Canada has been provided by private companies rather than government agencies. These countries were reluctant to subject these private enterprises to international regulation, and while they sent observers to conferences and meetings, they did not agree to be bound by any of the ITU's regulations until after the Second World War.

In 1947, the ITU voted to become a specialized agency of the United Nations (UN). It modified its structure and procedures to make them consistent with those of the UN and moved its headquarters from Berne to Geneva.

More recently, the ITU has moved to accommodate the interests of developing nations, which now are a majority of the organization's membership. One such accommodation was the reservation of satellite orbital positions and associated radio frequencies for use by countries that were not immediately able to take advantage of those resources. Another such accommodation is the addition to the ITU's charter of a technical assistance mission.

In 1992 the ITU approved a fundamental reorganization designed to make it more responsive to the rapid pace of technological change in international communications. This reorganized ITU was born officially on July 1, 1994, and is described in the following section.

2.2.2 The Structure and Procedures of the ITU

The ITU's highest authority is its Constitution, supplemented by a document called the Convention. Together, these two documents set out the mission, structure, and working methods of the ITU and define the rights and obligations of the organization's members. The provisions of both documents are binding on the ITU's members.

As the Constitution provides, the supreme authority of the ITU is the Plenipotentiary Conference, or "Plenipotentiary," which meets every four years. All member countries are represented at the Plenipotentiary, which sets long-term policy, elects officers, sets the budget, amends the Convention, and elects the 41 members of the Council. The Council (formerly called the Administrative Council) acts as the chief policy-making body between meetings of the Plenipotentiary. It meets once a year.

Day-to-day administration of the ITU is the job of the General Secretariat, headed by a secretary-general who is elected by the Plenipotentiary for a four-year term. The detailed work of the ITU is done at periodic conferences, each devoted to a particular area of the ITU's responsibility, and by permanent groups known as *sectors*, each of which is headed by a director elected by the Plenipotentiary. Each of these bodies is worth describing in detail.

ITU oversight of radio matters is delegated to a permanent Radiocommunication Sector (roughly corresponding to the former CCIR and now called the ITU-R), supervised by a World Radiocommunication Conference (WRC) that meets every two years to adopt revisions to the ITU's Radio Regulations [17]. (Regional radiocommunication conferences will convene as needed to consider issues associated with particular geographic areas.) The WRC also organizes and oversees the work of the study groups that deal with ongoing technical questions posed by RF spectrum use (for both terrestrial and satellite communications), radio system performance, and radio-related emergency and safety questions [18]. Administration of the Radiocommunication Sector is entrusted to the Radiocommunication Bureau, which is headed by a director.

ITU oversight of telecommunications standards (including the interconnection of radio systems used for telecommunications) is delegated to the Telecommunication Standardization Sector (roughly corresponding to the former CCITT, and now called the ITU-T), which is supervised by a World Telecommunication Standardization Conference that meets every four years. Like the Radiocommunication Sector, the Telecommunication Standardization Sector includes a number of study groups that work on technical problems associated with the performance and compatibility of telecommunications systems and equipment. This sector also includes a permanent Telecommunication Standardization Bureau, which is headed by a director elected by the Plenipotentiary.

The ITU's mandate to assist in international development is the province of the Telecommunication Development Sector (ITU-D), which is supervised by a World Development Conference that meets every four years. (Regional development conferences are also held as needed.) Like the other sectors, the Telecommunication Development Sector works through a number of study groups. It also includes a bureau, which is headed by a director, and an advisory board.

Within the Radiocommunication Sector is a part-time body called the Radio Regulations Board (RRB). The RRB is composed of nine members elected by the Plenipotentiary and is empowered to interpret the Radio Regulations and establish rules for the registration of frequency assignments. The RRB is supported by the Radiocommunication Bureau in such routine tasks as entering frequencies in the Master International Frequency Register.

Aside from the provisions of the Constitution and the Convention, the ITU enacts rules known as the Administrative Regulations. Some of these regulations—specifically, the Radio Regulations—are treaty undertakings that bind all member governments. (In ITU parlance, they are *Final Acts* of the organization, meaning that they have equal dignity with the provisions of the Constitution and the Convention.) Others—notably the telecommunications standards and development regulations—are nonbinding recommendations.

Telecommunications standards ordinarily originate with the work of the study groups, which make recommendations that are taken up at the quadrennial World Telecommunication Standardization Conferences. (If a study group decides that a recommendation is especially urgent, however, the recommendation can be adopted between conferences by correspondence with all ITU members.) The members have three months to approve or disapprove, and a simple majority will enact the recommendation.

Radio Regulations, unlike the telecommunications standards, are classified as Final Acts of the ITU when adopted at the Plenipotentiary. These regulations include the Table of Frequency Allocations, which specifies the lawful uses for all RF bands between 9 kHz and 400 GHz, the assignments of orbital positions and radio frequencies for satellites, and the rules governing the allocation of frequencies.

Because the ITU is a treaty organization, the delegations that work with its sectors and advisory bodies are accredited by the governments of the member countries. Where member countries' telecommunications networks are run by PTTs, the delegations are made up of civil servants. Where private companies provide these services, delegations may include employees of those companies who are accredited by their governments to participate. Finally, private companies may join the ITU as nonvoting members sponsored by their governments.

2.2.3 International Standards Organization

Another important source of telecommunications standards is the International Standards Organization (ISO), which was founded in 1946. The ISO has a broad mandate, ranging from agriculture to the garment industry to computers and computer communications.

The ISO is made up of a combination of *full members* and *correspondent members*. The former are national standards-setting organizations, while the latter are delegations from the governments of countries that do not have standards-setting organizations. Like the ITU, the ISO is an agency of the UN; unlike the ITU, the ISO is a nontreaty organization [19].

The ISO divides its work among plenary sessions of the organization (meeting every three years), a Central Secretariat in Geneva, and over 100 standards-making committees supported by some 2,000 subcommittees and working groups.

The most important of the ISO technical committees (TC), from the standpoint of telecommunications, is TC 97—the committee that develops standards for data processing and communications. Within TC 97, the most important of the subcommittees (SC) are SC 6, which has responsibility for data networking, and SC 16, which deals specifically with Open Systems Interconnection (OSI) [20].

Standards are adopted by vote: each proposed standard, called a *draft international standard* (DIS), is put out for comment and then referred for a vote to all ISO members who participate in the work of the relevant technical committee. A vote of 75% of the participating members is sufficient for the ISO council to adopt the DIS as an international standard.

The ISO's work, of course, overlaps to some degree with that of the ITU and many other standards-making organizations. The ISO coordinates its work formally with approximately 400 such organizations [21].

2.2.4 International Electrotechnical Commission

The International Electrotechnical Commission (IEC), like the OSI, is a voluntary, nontreaty international standards-making organization. The IEC concerns itself with electrical and electronic equipment standards not covered by the OSI [22].

Forty-four countries are IEC members, and each of those members has a counterpart, domestic committee to advance the development and implementation of IEC standards within its borders. Like the ITU and the OSI, the IEC divides its work among plenary sessions (a Plenary Assembly meets every year), a General Secretariat in Geneva, and a number of technical committees and subcommittees. Of particular interest to telecommunications is IEC TC 46, which sets standards for a number of telecommunications transmission facilities, including optical fiber. These activities involve considerable coordination with the ITU [23].

2.2.5 Some Regional and National Standards-Making Bodies

In addition to the international organizations just described, which are open to members from around the world and seek to prescribe standards of the broadest possible application, some other bodies prescribe standards of national or regional scope. We briefly describe some of the more important of these organizations here.

Conference of European Post and Telecommunications Administrations

Telecommunications service in European countries is often provided by government agencies known generically as PTTs. In 1959, the European PTTs formed their own organization designed to harmonize their jointly provided services, including mail delivery and telephone and telegraph services. A plenary session (the Plenary Assembly) meets approximately every two years, and a permanent staff in Bern (called the Liaison Office) handles routine administration between meetings of the Plenary Assembly [24].

The Telecommunication Commission of the Conference of European Post and Telecommunications Administrations (CEPT) is supported by a number of committees that work on the economic and technical coordination of PTT services within Europe and between Europe and other continents. These committees include a Coordination Committee for Satellite Communications, a Liaison Committee for Transatlantic Telecommunications, and a Coordination Committee on Harmonization.

The work of the CEPT has increasingly come under the jurisdiction of the EU. At the EU's request, the CEPT has developed a group of specifications that are binding on its members. These standards are known as NETs, which stands for "Normes Europeeans des Telecommunications," or European Telecommunications Standards. An organization known as the Technical Recommendations Application Committee (TRAC) selects certain standards developed by CEPT committees and the Telecommunication Commission and adopts those standards as binding NETs.

Since the creation of ETSI (described in the next section), the influence of CEPT in standards making has declined substantially.

European Telecommunications Standards Institute

The European Telecommunications Standards Institute (ETSI), founded in 1988, is likely to become the dominant European standards-making body. ETSI's membership is open to any European nation—not just members of the EU. ETSI also has categories of membership for non-PTTs, including private service providers, manufacturers, and user groups. It abandons any unanimity requirement in favor of a majority vote method of governance, with members' votes weighted according to their importance in European telecommunications.

Like the ITU, ETSI works through a large number of technical committees and subcommittees. The work of these organizations is supplemented by project teams made up of paid experts.

Standards produced by ETSI are known as European Telecommunications Standards (ETS). These may be preceded by Interim European Telecommunications Standards (I-ETS), which expire in two years unless they become ETSs or are renewed. To ensure coordination with the work of the ITU, ETSI is also a member of ITU-T.

Committee T1 of the Alliance for Telecommunications Industry Solutions

In August 1983, on the eve of the breakup of the Bell System, the Exchange Carriers Standards Association—now called the Alliance for Telecommunications Industry Solutions (ATIS)—recommended to the FCC that an ANSI-style committee be established to develop telecommunications standards for the postdivestiture industry. T1 has open membership and follows the ANSI requirements for procedural fairness to all participants.

The T1 Committee also supports the U.S. State Department's liaison functions with the ITU and other international standards-making organizations.

Bellcore

Bell Communications Research (Bellcore) was created at the divestiture of AT&T to assist the newly independent BOCs with technical support in the design and operation of their networks. Because Bellcore is owned by the BOCs and subject to the AT&T consent decree, it may not design or build telecommunications equipment.

The Bell companies' need for a separate, dedicated organization for joint network planning has not been as great as anticipated, and the Bells have announced their intention to divest themselves of Bellcore. While Bellcore is working actively to redefine itself as a private research and development group, the future direction of the organization is uncertain.

In the following sections, we consider particular standards that affect international telecommunications, both terrestrial and space-based.

2.3 ENSURING COMPATIBILITY: TERRESTRIAL TELECOMMUNICATIONS STANDARDS

In the early days of telecommunications, switching was less automatic and transmissions were analog. In this environment, most international standards were intended to ensure minimum transmission quality, and less coordination was needed to set up, signal, take down, and carry the call.

As telecommunications became more automatic and relied less on human intervention, technical standards became far more important. Today, the ability to complete transmissions across borders depends on achieving compatibility among different kinds of terminal equipment and private and public networks. We examine some of these standards here, beginning with the international direct distance dialing plan.

2.3.1 International Direct Distance Dialing

One of the most remarkable features of modern telecommunications is the ability to dial an MTS call almost anywhere in the world without the intervention of an operator. The advent of this capability required advances in the capacity and sophistication of switching and transmission technology. It also required that every telephone customer have a unique number in a format that any originating telephone switch could recognize.

The CCITT (now the Telecommunication Standardization Sector) met this need in 1964, when it adopted an international direct distance dialing (IDDD) plan based on numbers with a maximum of 12 digits [25].

The first two or three digits of an IDDD number are the country code. The first digit of a country code identifies the world zone in which the country is located (North America, for example, is world zone 1), and the remaining one or two digits represent the destination country [26]. The rest of the IDDD number is simply the number assigned to the called party by his or her domestic telephone company.

In addition to the present system of international numbering, the ITU has adopted an ISDN numbering plan that will take effect on December 31, 1996. ISDN numbers will have a maximum of 15 digits.

2.3.2 Data Communication Standards

When people talk over voice-grade MTS circuits, the network provides a channel of somewhat inconsistent, but generally adequate, transmission quality. Once the channel is established, the parties themselves manage the exchange of information between them in a loose, improvised fashion: they may say "hello" and "good-bye" to signal the start and end of the conversation, and each will (we hope) wait for the other to finish a thought before responding. When machines talk to each other in streams of binary data, the voice environment, with its loose conversational rules and high tolerance for error, may no longer be adequate. In some cases, the two-wire voice access line must be replaced with a higher quality channel or conditioned by error-correction methods, and the informal rules that guide human conversation must give way to complex, inflexible rules and specifications defined in software protocols and hardware interfaces.

The pervasiveness of equipment interfaces and protocols in data communications can be illustrated by an ordinary e-mail message. Suppose, for example, that your company subscribes to an e-mail service called AlphaMail, offered to the public by the AlphaNet Company. Using AlphaMail, you can exchange e-mail messages on a dialup basis with any of your company's employees at any of its far-flung locations. AlphaMail also offers interconnection with a private e-mail service operated by BetaNet, one of your company's principal suppliers, which runs an internal e-mail service over its private packet data network.

You decide to send an e-mail message to your friend Joseph at BetaNet. You select the communications software package on your desktop PC and go to AlphaMail. When the modem has dialed AlphaMail's number and the connection is established, you type Joseph's e-mail address—"joseph@beta.com"—and a message.

Your computer converts the address and message into a stream of bits, and the modem converts those bits into a series of analog tones for transmission over the public switched telephone network. The analog tones travel to AlphaNet's packet data network, where a second modem demodulates the tones into a digital bit stream.

Your message's journey has barely begun, but already you have relied on some communication standards. For one thing, because you do not have a four-wire private line, you have selected a modem with dialing capability, as well as the ability to perform error correction and echo suppression to compensate for the inconsistent quality of dialup circuits. Unless the modem incorporates these features, the interface between the PC and the network will not be established or may fail because of transmission errors. Similarly, your modem must be technically compatible with the receiving modem at AlphaMail's facility: for example, if your modem uses asynchronous transmission, the receiving modem must use the same method or the AlphaMail service must perform a conversion between the two modems [27].

We assume that the modems have done their job and your message has arrived intact at AlphaNet. The service now assembles an *envelope* around the e-mail message, which may include a message identifier, the originator's name, the recipient's name, the names of anyone to be copied on the message, the subject, content type, and level of confidentiality. The language and format of this envelope are also standardized: if the envelope protocol used by AlphaMail was inconsistent with the one used by the BetaNet system, the message would not arrive or would not reach the intended recipients.

AlphaNet next prepares the message, with its e-mail envelope, for transmission over AlphaNet's packet-switched network. It breaks the bit stream into separate packets and augments each packet with information identifying the transmission to which the packet belongs, the order in which it will appear in the reassembled transmission, and the address on the packet network to which it must be delivered. This process, too, is controlled by protocols: the AlphaMail system must use the same routing protocol as the BetaNet system; otherwise, the packets will arrive out of order (making the reassembled message unintelligible) or in the wrong place.

A high-speed switch now sends the packets through a series of nodes along transmission paths shared with packets from many other transmissions. (Each node checks the packets for errors and corrects the errors or directs the origination point to retransmit.) The packets are eventually handed off to the BetaNet packet-switched network, where they are reassembled and delivered to the BetaNet e-mail service. The e-mail system polls Joseph's PC to determine whether it is ready to receive a message; if not, the message is stored and will be forwarded to Joseph's computer when he requests it.

As our example shows, an apparently simple data communication may require the setting up, managing, and taking down of many separate data transactions. Each of those transactions is a cooperative process among at least two corresponding networks or pieces of equipment, controlled by precise interfaces and instructions. Unless each of these pairs of cooperating entities speaks the same language, the entire complex structure of the e-mail message will fail.

Unfortunately, compatibility among networks and equipment has often been the exception rather than the rule. When incompatible protocols are used, LANs may not be able to access other LANs or interconnect with the public data network; customers using one packet-switching network may be unable to reach services served by another packet-switching network, and customers of one e-mail service may be unable to correspond with customers of another e-mail service.

In the days when data communication took place mostly within large organizations that could choose to deal with a single network vendor, compatibility problems arose less frequently and had little impact beyond the affected organization. As data communication has become pervasive, however, the demand for ubiquitous interconnection has become irresistible. Even network vendors with proprietary protocols understand that the more users their networks can reach, the more valuable their services become.

International standards-making organizations (particularly the ITU and ISO) have responded to this demand by developing a number of *open* standards for the design of data communication protocols. To the extent network and equipment providers adopt these standards instead of *closed*, proprietary protocols, the goal of universal interoperability will be achieved.

Central to this effort is the ISO's Reference Model for Open Systems Interconnection (the OSI model), which defines a structure within which open standards can be defined.

The Reference Model for Open Systems Interconnection

The OSI model is hierarchical; each data transmission is portrayed as a set of parallel activities, with each activity dependent on the one "below" it and supporting the one "above" it. One way to understand this hierarchy is to compare it with the division of computer software into operating systems, which provide a transparent platform on which applications can run, and application systems that make use of the operating systems to perform particular tasks. If we imagine our computer divided into seven layers, with the hardware as the first layer and each layer of software providing a kind of operating system for the one above it in the hier- archy, then we can imagine something like the OSI model.

Another way to approach the OSI concept is to return to our e-mail message. Specifically, notice how some of the interfaces and protocols we described were specific to the handling of e-mail, while others were independent of that application. Notably, neither the modems nor the packet networks "know" that the information encoded in the bit streams is an e-mail message—it could as easily be a request for a file, a directory lookup, or a signal controlling a voice telephone call. Those systems are concerned only with moving the bit stream in an error-free fashion. Other protocols—such as the format of the e-mail envelope and the method of storing messages in the recipient's mailbox—are peculiar to messaging services generally or to e-mail in particular. They make use of the less specific protocols that ensure error-free bit transmission, but do not themselves "know" how those protocols do their jobs (just as the parties to a voice conversation are not conscious of the transmission methods by which their voices are carried over the telephone network).

In the OSI model, the lower level functions (those that are independent of the particular task in which the end users are engaged) are referred to as *network* functions. The higher level activities (those that use the network functions to perform specific tasks) are called the *end-to-end*, or *end-user*, functions.

The first layer of the model, and the foundation for the network functions, is the *physical* layer. This is the interface standard or protocol that defines the physical, electronic movement of the bit stream. In our e-mail example, some of the standards governing operation of your modem were physical standards.

The second layer of the model is the *data link* layer. Protocols at this level assume that the bit stream has been generated and correct errors—such as crosstalk, current surges, or interference—that occur between nodes on the transmission path. We saw a standard of this kind in our e-mail example when we discussed error correction on the link between the two modems. The method of addressing characters on an asynchronous modem-to-modem link can also be handled at the data link level.

The third layer of the model is the *network* layer. Protocols at this layer assume that the bit stream is moving across a link (layer 1) and that the transmission is error-free (level 2). The network protocols take care of the setup, control, and release of the link connection. In our e-mail example, the packet-switching protocols were network layer standards. Those protocols governed the routing instructions for movement of packets around the network and set up and took down the connection needed to move data from the AlphaNet packet network to the BetaNet packet network.

Once we advance beyond the three network layers, we deal with protocols that govern communication between end users, rather than intermediate links in the transmission chain. So, in our e-mail example, network protocols established, managed, and took down the modem-to-modem connections, the connections within the AlphaNet and BetaNet networks, and the connections between those networks. The protocols above layer 3 are independent of these intermediate links.

The first of the end-to-end layers is layer 4, the *transport* layer. This is the highest layer that concerns itself with the transport medium and the lowest layer that is not confined to a particular link in the transmission chain. It may include programs to pick the cheapest carrier or routing for a transmission. It might also decide whether a message should be multiplexed and then manage the multiplexing process. The transport layer might be viewed as an efficiency expert, managing the work of the network layers.

Layer 5, the *session* layer, functions somewhat like the network layer—except that unlike the network layer, the session layer concerns itself with end-to-end communication between the users, rather than with the links and nodes. This is the layer that determines whether the users will have a two-way conversation, whether a message will be delivered, stored and forwarded, or whether information will be requested and retrieved from a database. The session layer will manage the transmissions accordingly.

Layer 6, the *presentation* layer, performs changes in syntax between end-user devices. Functions performed at this layer include encryption and data compression.

Finally, Layer 7 is the *application* layer. Protocols at this layer perform services such as file transfer, message handling, and directory search. These functions are specific to particular applications (such as e-mail), but are not themselves the end user's application software. The application software is above this layer and works directly with it.

While a complete exposition of the OSI model is beyond the scope of this book, some of the more notable standards already developed within the OSI architecture are worth describing.

X.400: The Message-Handling (E-Mail) System

The ITU has adopted a set of standards, functioning at the upper layers of the OSI model, for interoperability among electronic messaging services. X.400 divides e-mail networks into user agents and message transfer agents. The user agent takes the message from the sender and prepares the envelope we described earlier. The message transfer agent transmits the message through the system, typically over a packet-switched network. The standard also specifies a protocol for communication between user agents, another protocol for communication between a user agent and a message transfer agent, and a third protocol for communication between message transfer agents.

The emergence of e-mail as a ubiquitous service depends heavily on the interoperability of e-mail services. Industry acceptance of X.400 has so far been encouraging.

X.500: The Directory Standard

The ITU has worked for several years on the development of protocols to support a global directory of telecommunications users. The directory, codified as X.500, is perhaps the most complex communication protocol ever attempted.

The directory protocol specifies a number of *address* and *attribute* categories and a hierarchical information tree for structuring searches. It also specifies a system structure consisting of *directory user agents* (DUA), which request and retrieve data at the end user's request, and *directory service agents* (DSA), each of which contains a portion of the database and can query other DSAs as needed.

While the X.500 protocol is unlikely to produce a single, global directory, it provides a structure within which directories of more limited scale can be implemented and interconnected. X.500 also provides protocols for the security of directory information.

X.25: The Packet-Switching Standard

The X.25 standard defines the interface with the packet data network at the network, data link, and physical layers. This is perhaps the most successful of the data communication standards, with almost universal support among providers of packet-switching services.

X.25 defines a *connection-oriented* service, meaning that a path for transmission of the packets (a virtual circuit) must be established before the packets are sent through the system. X.25 also calls for substantial error-correction activity at various nodes along the transmission path. These features distinguish X.25 from services such as frame relay (which does not perform error correction) and SMDS and LAN protocols that work in a connectionless mode. Because of these differences, some of the newer data-switching technologies are both faster and more efficient than X.25.

The Fax Machine Standards

The ITU has implemented a series of equipment standards for fax machines. The Group 1 and Group 2 (Recommendations T.2 and T.3) standards specify analog machines using amplitude modulation and frequency modulation, respectively, and taking six minutes and three minutes, respectively, to scan and transmit a page of text.

The Group 3 standard (Recommendation T.4) was the first specification for a digital machine. Group 3 machines have high digital resolution quality, transmit a page in 20 seconds or less, and send data over the voice telephone network using a modem. Group 3 machines also send back a confirmation that can be printed by the originating machine.

Group 3 machines achieve their high rate of transmission, in spite of the limitations of the public telephone network, by compressing data and eliminating redundancy. Notably, although each line is scanned as a series of equal-sized units called *pels*, long stretches of white are not transmitted as individual pels but are simply summarized as a pel count.

Group 4 (Recommendations T.5 and T.6) machines provide better copy quality than Group 3 machines at higher speed, but must be used with dedicated data lines rather than voice-grade telephone lines. Accordingly, acceptance of Group 4 machines by a mass market probably must await the deployment of ISDN or other high-speed digital facilities to residential and small business telephone customers.

Transport Control Protocol/Internet Protocol

Transport control protocol/Internet protocol (TCP/IP) is a packet-switching protocol developed by the U.S. Department of Defense. It has proved popular with universities and drives the system of interacted packet networks known as the Internet, which is of course enjoying exponential growth.

While TCP/IP is not an OSI protocol, it operates approximately at the transport and network layers of the OSI model. Continuing development of the protocol and related standards is done by an organization called the Internet Engineering Task Force (IETF).

2.3.3 Standards for Connection to Public Networks

As we noted earlier, one of the most difficult challenges posed by privatization and competition in the telecommunications industry is the need to interconnect customer-provided equipment, value-added services, and competing carrier facilities to the dominant public network. As the FCC decided in the United States, the terms and conditions of interconnection cannot be left to the monopoly public networks to decide, but must be worked out through a combination of voluntary and de jure standards making.

The degree of governmental intervention in this process varies substantially from one country to the next. Some countries simply direct the dominant carriers to negotiate interconnection terms in good faith with their competitors, and will involve themselves in the process only to resolve complaints. Other countries have enacted statements of principles and in some cases detailed rules to guide the conduct of the parties.

While most of these efforts are part of the domestic regulatory regimes of individual countries [28], some multinational access rules have been enacted. Notably, the EU established a framework for ONP in 1990, calling for member PTTs to provide interconnection for value-added and other authorized facilities and services on tariffed terms at cost-based rates and subject to published, nondiscriminatory technical terms of interconnection. The EU subsequently issued a more specific directive in 1992, requiring member PTTs to provide defined categories of leased lines and interconnect with those facilities on a nondiscriminatory basis by 1993. Implementation of this directive has been uneven, but the EU remains committed to the ONP approach [29].

Notes

[1] We give some examples of these coordination tasks in Section 2.3.2.

[2] *De jure* standards include, for example, the equipment standards adopted by the FCC for connection of customer-provided equipment to the U.S. public telephone network. *Voluntary*, or *consensus*, standards include the many recommendations of the consultative committees of the ITU, some of which are discussed in this chapter. *De facto* standards include, for example, the VHS standard for video cassette recorders and the DOS and Windows operating systems used in IBM-compatible personal computers. See, for example, S. Besen and G. Saloner, "The Economics of Telecommunications Standards," in *Changing the Rules: Technological Change, International Competition, and Regulation in Communications*, R. Crandall and K. Flamm, eds., Brookings Institution, Washington, DC, 1989, pp. 178–179. See also G. Wallenstein, *Setting Global Telecommunications Standards*, Norwood, MA: Artech House, 1990, p. 15 et seq.; and P. Bernt and M. Weiss, *International Telecommunications*, Carmel, IN: Sams Publishing Co., 1993, pp. 122–123. See generally U.S. Congress, Office of Technology Assessment, *Global Standards: Building Blocks for the Future*, OTA-TCT-512, 1992.

[3] Standards that are formalized before their use in the marketplace are sometimes called *a priori* standards. For an extensive discussion of the a priori process, see Wallenstein [2], p. 19 et seq. We should also be aware that the same standard may be categorized differently at different places and times. So, for example, what began as a de facto standard may be converted by regulators into a de jure standard or be adopted by an international organization as a consensus standard, or a standard that is de jure in only one country may nonetheless be the unofficial, de facto standard in other countries.

[4] In a famous speech given in 1973, AT&T's chairman warned that "[a] faulty telephone in one house could conceivably disrupt service to an entire city." Quoted in S. Coll, *The Deal of the Century*, New York: Atheneum, 1986, p. 105. Unfortunately for AT&T, in a series of antitrust

proceedings extending over more than a decade, no evidence of an incident of this kind was ever found.

[5] In the terms we used earlier, AT&T had substituted its own de facto standard for the de jure standard the FCC might have developed or the voluntary standard AT&T might have worked out in cooperation with its competitors.

[6] 456 U.S. 556 (1982).

[7] In the Supreme Court, the principal question was whether ASME could be held responsible for the acts of its committee members, which those members had characterized as "unofficial." The Court concluded that such liability was appropriate. Ibid., p. 570.

[8] *Indian Head, Inc. v. Allied Tube & Conduit Corp.*, 817 F.2d 938, 941 (2d Cir. 1987).

[9] Allied Tube & Conduit Corp. v. Indian Head, Inc., 486 U.S. 492, 509 (1988).

[10] We describe the work of the ITU and its specialized committees in Section 2.2.

[11] Besen and Saloner [2], p. 215.

[12] Ibid.

[13] For an explanation of digital and analog modulation systems, see Chapter 1.

[14] Japan also adopted the North American standard.

[15] For an excellent, concise discussion of the U-interface problem, see J. Pecar, R. O'Connor, and D. Garbin, *Telecommunications Factbook*, Appendix A, 1993, pp. 323–325.

[16] See, for example, Wallenstein [2], p. 19 et seq.

[17] The WRCs replace the former World Administrative Radio Conferences (WARC).

[18] Immediate supervision of the study groups is entrusted to the Radiocommunication Assemblies, which convene every two years in support of the Radiocommunication Conferences.

[19] See Wallenstein [2], pp. 85–89.

[20] We say more about the ISO's Open Systems Interconnection model on pp.20–21.

[21] See, for example, Besen and Saloner [2], pp. 188–89.

[22] See Wallenstein [2], p. 84.

[23] Much of the ground covered by the ISO and IEC is also covered by two regional European organizations—CENELEC (Comite Europeen de Normalisation Electrotechnique) and CEN (Comite Europeen de Normalisation). These two organizations, founded in 1958 and 1961, respectively, impose uniform standards on all computer and computer terminal products sold in Europe. Wallenstein [2], pp. 57–58.

[24] See Wallenstein [2], p. 202.

[25] See, for example, B. Elbert, *International Telecommunication Management*, Norwood, MA: Artech House, 1990, p. 89–90; and Bernt and Weiss [2], p. 243.

[26] The first digit of the country code can be any number except 0, the second digit must be a 0 or 1, and the third digit may be any number between 0 and 9.

[27] When data are transmitted in a synchronous mode, the sending and receiving equipment uses a shared timer to govern the transfer of information between them. When data are exchanged in an asynchronous mode, each piece of equipment (in our example, each modem) uses start and stop signals rather than a shared timing mechanism to separate the transmitted characters.

[28] Examples are the *Computer III* rules, discussed elsewhere in this book, and the Japanese Open Network Development (OND) initiative.

[29] See, for example, "EC Approves Telecom Liberalization Directive," *Telecommunications Reports*, July 24, 1995, p. 27; and "Court Faults Greece Delay on Leased Line Directive," *Telecommunications Reports International*, July 21, 1995, p. 6.

Scarce Resources: Radio Spectrum and Orbital Positions

3

The physical resources needed for wireline electronic communication, such as transistors, switches, cables, optical fiber, and physical rights of way, are allocated among users by the marketplace. Generally speaking, regulatory bodies do not ration switches and cables or stamp those facilities with a list of the services for which they may be used.

The electromagnetic frequencies that are useful for radio communication, however, are different. They are scarce, and simultaneous transmissions on the same frequency in the same area will result in interruption or disruption of the competing signals—a phenomenon known as *interference*. In order for the RF spectrum to be used most beneficially, therefore, someone must deal with scarcity by rationing the spectrum among various applications and must prevent interference by granting some users the right to transmit on particular frequencies to the exclusion of others [1].

These two functions—usually referred to respectively as *allocation* and *allotment* of frequencies—are performed by both national and international regulators. Where radio transmissions will begin and end within the borders of a single country, allocation and allotment of the frequencies used for those transmissions may be a proper subject of domestic regulation. Where radio transmissions will cross borders, traverse extraterrestrial space or be sent and received by roving ships and aircraft, however, international coordination and regulation of those transmissions are needed. Orbital locations of satellites constitute a special instance of this RF management problem. As we discuss more thoroughly below, the orbits useful for geostationary communications satellites are concentrated in a narrow band 22,300 miles above the equator. Spacing between satellites occupying this band is critical—not because of the risk that the satellites will bump into each other (the chance of physical contact is slight), but because the radio transmissions between the satellites and points on the Earth's surface, when made on the same frequencies, will cause interference if they encounter each other [2]. Here, as with spectrum management generally, a combination of national and international regulation is needed to allocate the resource among services and allot transmission rights to individual users.

The discussion that follows describes the general scheme of international radio spectrum management and the particular problems of satellite orbit regulation. Both subjects, as we will see, are primarily the province of an organization we have encountered before: the ITU [3].

3.1 INTERNATIONAL RF MANAGEMENT

The ITU performs a number of functions related to management of the RF spectrum. It establishes technical standards to govern the power, modulation techniques, and other properties of radio emissions; it allocates frequencies to particular services; it makes allotments of frequencies among nations; and it defines conditions under which users of spectrum have the right to operate free of interference. The allocation and allotment processes, in particular, require some explanation.

When we speak of frequency allocations, we refer to the process by which the ITU identifies particular ranges of radio frequencies as appropriate for particular services. So, for example, the ITU has set aside certain frequency ranges for broadcasting, another for radar, another for maritime radio, another for satellite communications, another for radio astronomy, and so forth. Allocations are decided on at the World Radiocommunication Conferences (WRCs) and recorded in the Table of Frequency Allocations set out in the Radio Regulations. Once adopted as part of the Radio Regulations, the frequency allocations are binding upon ITU members. (Appendix A shows a portion of the Table of Frequency Allocations as incorporated in the rules of the U.S. Federal Communications Commission.)

This is not to say, however, that each member country of the ITU uses the radio spectrum in precisely the same way. The Regulations themselves often recognize more than one possible use for the same group of frequencies and sometimes allocate the same frequencies to different services in different regions of the world [4]. In addition to these variations, ITU members may record their intention to deviate from particular allocations by making reservations to the Regulations and placing footnotes in the Table of Frequency Allocations [5]. Not surprisingly, the allocations of the more popular frequency bands are littered with exceptions of these kinds.

Once frequency bands have been allocated to particular services, someone still must decide who may provide the prescribed services over the prescribed frequencies. Within nations, licensing of frequencies to particular users (called *assignments* in ITU parlance) are made by the national regulatory authorities. But before assignments may be made within nations, it is sometimes necessary to allot frequency bands within allocations to particular nations or regions. In order to understand this international allotment process, we must understand when it is necessary and the principal approaches that have been taken.

International allotments are necessary only when there is a possibility of international interference. So, for example, the signals of local commercial broadcast stations are normally confined within metropolitan areas, and terrestrial microwave transmissions cause no appreciable interference outside their line-of-sight transmission paths. Accordingly, national administrations may typically use the entire frequency band allocated to such a service, assigning it to their domestic licensees without advising or consulting the ITU [6]. But long-range broadcasts, satellite transmissions, and many other services cannot be confined within the borders of a single country. When licensees of a service in one country will have the potential to interfere with the signals of other nations' licensees, those nations cannot separately assign the entire frequency spectrum allocated to that service among domestic licensees: instead, those countries must find a way to share the available frequencies.

In practice, frequency allotment is the work of several organizations. The chief agent of global cooperation is, of course, the ITU; however, the ITU's work is supplemented by that of a number of groups that address regional interference problems, as well as bilateral arrangements that address more specific problems.

Whatever the organization or the scope of its work, international frequency management displays a fundamental tension between two approaches—the *a priori* and the *a posteriori* approaches. Simply put, a priori management allots spectrum among users according to a plan, while a posteriori management (sometimes called *first come, first served*) merely establishes the circumstances under which users' claims to spectrum will be recognized. Under the first scheme, rights are vested at the initiative of the planning organization; under the second, rights are vested at the initiative of those who claim them. Not surprisingly, a posteriori methods tend to be popular with the larger, more entrenched users of radio services and less popular with small or developing countries that fear the exhaustion of spectrum before they are in a position to use it [7].

Over many years, the ITU has met with mixed success in its efforts to implement a priori methods of assigning rights to spectrum. Notably, in 1947 the ITU, at its Atlantic City Administrative Radio Conference, created an approach called the *engineered spectrum*. The idea was to review all existing frequency uses and develop a plan for the equitable allotment of spectrum among nations. In order to administer the engineered spectrum, the ITU created the International Frequency Registration Board (IFRB).

Among other functions, the new IFRB would maintain a Frequency Register (now known as the Master International Frequency Register) of all frequency assignments. Reported assignments that conformed to the ITU's frequency allotments would be recorded in the Registration (2a) column and would enjoy maximum protection from interference. Nonconforming assignments that did not otherwise violate the Convention and Regulations and did not cause interference with other assignments were placed in the Notification (2b) column. Assignments that did not conform with the allotment plan and caused harmful interference would not be registered.

The ITU's engineered spectrum was implemented only for a limited number of services, and the ITU eventually adopted a column 2d for recordation of frequencies as to which no allotment plan had been devised [8]. Column 2d assignments now predominate and are at the center of the ITU's a posteriori, notice-and-recordation procedure.

Under the present notice and recordation system, countries that intend to use particular sets of frequencies notify the IFRB of their intentions. (While this sounds simple, the disclosures required by the IFRB are quite complex and vary among services and frequency bands.) The IFRB then determines whether the registration complies with its regulations and presents a threat of interference with existing, lawful users of the frequencies in question [9]. If the registration passes muster, it is entered in the Master International Frequency Register and the registrant acquires priority against other users who may wish to transmit interfering signals over those frequencies.

While the IFRB's procedures are complex and impressive, the real-world effect of the notice-and-recordation procedure is uncertain. The IFRB process is meant to enforce the principle, set out at the ITU convention, that all radio stations must operate in such a manner as not to cause interference, and the concept of notification by date gives early registrants the high ground in any dispute with later, interfering registrants (at least as long as the early registrants comply with the Convention and Regulations in their use of spectrum). But the IFRB cannot act against an assignment that it finds to be contrary to the Convention or Regulations or that presents a risk of interference. (In fact, the IFRB cannot even refuse to record such a nonconforming use on the Master International Frequency Register.)

While the present procedures fall substantially short of the engineered-spectrum ideal, the fact is that most nations find it in their interest to resolve interference problems cooperatively, and the IFRB's recording and mediation activities are a useful part of this process. Whatever competing interests nations may have in the use of spectrum, nature has arranged it so that neither the interfering user nor the interfered-with user can send clear signals unless they find a way to share the resource.

3.2 MANAGING SPECTRUM AND ORBITS USED BY GEOSYNCHRONOUS SATELLITES

As we pointed out in the first chapter of this book, communications satellites can achieve continuous coverage of a large area when they occupy a fixed point above the Earth—in other words, when they are in geostationary orbit (GSO). A satellite in GSO can send radio signals to as much as 40% of the Earth's surface, and because it does not move relative to the surface of the Earth, it does not have to be "tracked" by the Earth stations with which it communicates. In fact, communications between geostationary satellites and their associated Earth stations are simple point-to-point microwave transmissions.

GSO is a scarce resource. Satellites will maintain a fixed *station* over the Earth only if they are launched into a band 22,300 miles above the equator. Once in GSO, satellites use navigational aids and small onboard rockets to correct a natural tendency to drift and to maintain position within a range of about 30 km in altitude and within about 0.1 degree of latitude and longitude. (And, of course, each satellite must occupy a longitude within GSO from which it can communicate with the part of the Earth it is meant to serve.) The result is a relatively narrow highway, particularly in the more popular segments of the orbit [10].

The scarcity problem is aggravated by the fact that communications satellites are radio stations, and like all radio stations, they must avoid interference with other stations. The problems posed by spectrum scarcity in space are both technical and legal, and each set of problems is worth reviewing in turn.

3.2.1 Technical Problems of Geostationary Orbit

The heart of any communications satellite is its complement of onboard transponders. Transponders receive signals (uplinks) from Earth stations and retransmit those signals (downlink) to other Earth stations. In order to keep the satellite from interfering with its own transmissions, the uplink and downlink are transmitted at different frequencies.

Most communications satellites now operating use the C band, which is also used by terrestrial microwave facilities. C-band satellites receive uplink signals in the 6-GHz range and transmit downlinks in the 4-GHz range. A more recent—and rapidly growing—generation of satellites uses the K band, a higher frequency range that does not interfere with terrestrial microwave. The first group of K-band satellites to see service, called the *Ku-band group*, receives signals at 14 GHz and transmits at 12 GHz [11]. Satellites of either class pose the same frequency coordination problem: they must be spaced far enough apart so that their signals do not encounter each other. Specifically, uplinks intended for one satellite must not be received by an adjacent satellite, and downlinks intended for one Earth station must not be received by another Earth station.

Avoiding interference among satellites is accomplished primarily by orbital spacing [12]. C-band satellites, for example, are typically spaced between 3 and 5 degrees apart, which works out to an interval of 2,200 to 3,600 km [13]. More recently, improvements in antenna design have permitted spacing of as little as 2 degrees, or 1,500 km, between satellites using the same frequency bands [14].

As antenna designs improve and additional frequency bands are made available for satellite service, the problem of orbital crowding may become less severe. For now, however, the allotment of desirable orbital locations within the available frequencies is a highly charged problem of international regulation. The next section offers a brief survey of the history and present status of these issues.

3.2.2 Regulatory Problems of Geostationary Orbit

When the USSR launched the first artificial Earth satellite in 1957, the rights of nations to use space for scientific, military, and commercial ends were undefined. Notably, no recognized categories of space-based services had been established, no radio frequencies had been allocated to space applications, and no principles or procedures for coordinating the use of spectrum and orbital positions among nations operating satellites had been announced [15]. Because Earth satellites used the RF spectrum in competition with other uses, these problems fell naturally within the competence of the ITU.

While the complete history of ITU regulation of the GSO is well beyond the scope of this chapter, we can sketch a broad outline of that history. At the risk of some oversimplification, we divide the story into four phases.

The Formative Phase: 1959 to 1977

The ITU first addressed space communication at the 1959 World Administrative Radio Conference (WARC), where it established a definitional framework for space regulation, added two space-based services to the Radio Regulations, and allocated spectrum for use in space research [16]. The 1959 WARC conferees recognized that their work accommodated existing uses, but did not provide for future needs, and they accordingly recommended that a special administrative conference be held to fashion a more satisfactory regulatory framework.

The special space conference, called the Extraordinary Administrative Radio Conference (EARC), convened in 1963. The EARC defined two broad categories of space services used for communication, allocated spectrum to one of those services, and devised a new coordination procedure for use of spectrum by satellites and Earth stations.

In defining the new satellite services, the EARC recognized that satellites used for communication between points on Earth (as opposed to research) would be used both for point-to-point communications (e.g., telephone calls) and for distribution of broadcast signals to the public. The conference classified these two kinds of service, respectively, as the communication-satellite service and the broadcast-satellite service.

The 1963 conference also allocated frequency bands to the communication-satellite service (but not to the broadcast-satellite service). Specifically, the conference allocated 2,800 MHz of spectrum to communication-satellite services, of which only 100 MHz was available for that use exclusively. The remaining 2,700 MHz was to be shared with terrestrial microwave services [17]. This sharing approach necessitated a complex coordination procedure for the use of the spectrum allocated to communication-satellite service. Under the new procedure, codified in Article 9A of the Regulations, any country intending to operate an Earth station [18] on the frequencies shared with terrestrial microwave services had to complete a coordination procedure *before* it sought registration with the IFRB. The country planning to operate the Earth station had to send detailed notice of its intentions to every country within "coordination distance" of the proposed station—that is, every country within the distance at which the station's signal might cause interference. Any country thus notified then had 60 days in which to agree or disagree with the notifying country's proposal. In the event of nonresponse, disagreement between the parties, or other problems, the IFRB could assist with the process upon request. As with all of the IFRB notice-and-recordation procedures, however, no judicial or enforcement mechanisms were available to resolve disputes that did not yield to a voluntary or mediated solution. In fact, if the coordination process was properly followed, the IFRB was obliged to register the Earth station whether coordination had succeeded or not [19].

The next major ITU conference devoted exclusively to space was the World Administrative Radio Conference for Space Telecommunications (WARC-ST), held in Geneva in the summer of 1971. This conference, based on years of preparatory work by the CCIR and several of its study groups, revised the space service definitions, allocated more spectrum, changed the coordination and registration procedures, and (perhaps most important of all) began the process of long-range planning for use of the GSO.

On the definitional side, the 1971 conference replaced the communication-satellite service, first defined in 1963, with two new services: the fixed-satellite service (FSS) and the mobile-satellite service (MSS). (The broadcast-satellite service definition stayed essentially the same as that adopted in 1963.) On the frequency allocation side, the 1971 WARC allocated about 15,000 MHz of new spectrum for satellite services, including spectrum for BSS. As with the 1963 allocations, many of the bands allocated were shared with terrestrial services [20].

As for the question of system coordination, the 1971 WARC adopted a procedure for mandatory notification of intent to establish satellite systems, as well as Earth stations. Article 9A of the Regulations required any administration planning to establish a satellite system to send certain information to the IFRB not earlier than five years before operation. The information furnished was to include the orbital position of all proposed space stations and the frequencies and power outputs at which the stations would operate. If the information furnished showed a risk of interference with any other administration (country), then coordination with the affected administration(s) was required.

The notice-and-recordation procedure fell short of detailed regulation of the GSO, and many developing countries believed that the time had come to move toward some kind of a priori plan that would guarantee their access to the orbit. This concern was addressed, if not resolved, by Resolution Spa 2-1, which affirmed that all countries enjoyed equal rights in both the frequencies and orbital positions available for geostationary satellite operation, and that registration of frequency assignments with the ITU *did not*, accordingly, give the registrant a permanent priority over future claimants to the resource. (The effect of this disclaimer of perpetual preference was, of course, uncertain.)

The WARC-ST also considered, for the first time, the vexing problems posed by DBS—a subset of the BSS that raised the possibility of high-power satellite transmissions reaching small, inconspicuous receive-only dishes located at individual residences. While no such system was in operation in 1971, Eastern Bloc and third-world countries were already concerned about the potential of DBS to bypass official censorship. These countries therefore urged a right of *prior consent*, or at least an a priori frequency allotment plan for the BSS that would recognize their concerns. The WARC-ST did not resolve this issue, and in fact declined to apply its coordination procedures to BSS. Instead, the WARC-ST adopted a resolution that could be interpreted (if one wished) as a nod in the direction of prior consent, called vaguely for a priori planning of spectrum use for BSS, and established for the interim a special coordination procedure for the service.

In 1973, the ITU's Plenipotentiary Conference at Málaga-Torremolinos made two important contributions to international satellite service regulation. First, the Plenopotentiary lent considerable impetus to the ideal of equal access to space resources by incorporating the principles of Spa 2-1—declaring the GSO to be a scarce resource to which all countries should have equal access—in the ITU Convention. Second, the Convention was amended to augment the IFRB's regulatory powers, including the recordation of orbital positions as well as frequency assignments. Finally, the 1973 Plenopotentiary addressed the unsettled issue of allocation of frequencies to the broadcast-satellite service by announcing a WARC to be held not later than April 1977. The resulting conference—the 1977 WARC for the Planning of the Broadcasting-Satellite Service—was the first real effort to allocate space communication resources on an a priori basis and was the beginning of a highly politicized phase of ITU regulation.

The Political Phase: 1977 to 1979

As we noted in the previous section, as early as 1971 the ITU recognized the equal right of all countries to positions and frequencies in the GSO, and declared in principle that first-come, first-served registration did not confer permanent rights. Developing countries, which now were a majority of the ITU's membership, took this principle very seriously: they saw satellite communications as a shortcut to the establishment of modern communications networks [21], and they did not want the developed countries to occupy all the better orbital locations before other countries were ready to use them. By the time of the 1977 conference on the BSS, therefore, the developing countries were ready to demand a serious effort at a priori planning.

The stated purpose of the 1977 conference, held for five weeks in Geneva, was to ration the 12-GHz band for satellite broadcasting on a basis that would ensure equitable access for all countries, large and small, whether or not represented at the conference. The resulting plan of allotment was highly equitable, but also unfortunately rigid and inefficient.

The 1977 plan applied to the use of satellites for domestic broadcasting in Regions 1 and 3. Planning for Region 2 (Western Hemisphere) was deferred because of the opposition of the United States to the plan [22]. The plan divided the GSO for these regions into positions with 6 degrees of separation and allocated downlink frequencies for each orbital position. (Planning for uplink allocations was deferred to the next WARC, scheduled for 1979.) The orbital positions, frequencies, and coverage areas were allotted to particular countries according to the geographic area they covered.

While the 1977 BSS plan was a victory for the developing countries, it has been a source of frustration for other countries seeking to operate satellites to serve Regions 1 and 2. Notably, the 1977 plan gives priority to countries that may lack the current ability or intention to utilize a spectrum-orbit allotment, even though other countries are ready to use it and will employ it more efficiently [23]. Also, the plan's requirement for 6 degrees of separation has been overtaken by improved technology, and the plan's country-specific assignments have retarded the development of multinational satellite broadcasting systems. These problems were predicted by the United States, and were largely responsible for its refusal to support the plan for Region 2.

In spite of the rigidity of country-specific, a priori allotments of specific orbital positions, the developing countries were determined to expand that approach at the General WARC held in 1979 [24]. By this time the developing countries were two-thirds of the ITU's membership, and they had had more experience with the difficulties of coordinating the use of orbital positions with INTELSAT and the space agencies of the developed countries [25]. The 1979 conference, therefore, included the airing of a number of so-called North-South disputes.

Some of the sharpest debates at the 1979 conference involved the FSS. Developing countries wished to have frequencies in the range below 10 GHz allocated to the FSS because the equipment needed to use these frequencies was less costly than higher frequency equipment. Frequencies in this range, however, were used extensively by terrestrial radio services operated by the U.S. and NATO military forces. The developing countries' solution was to assign a lower priority to radar applications in the range below 10 GHz—a proposal that caused a significant North-South dispute. The developed countries of the West, while eventually accepting a compromise on this question, had been forced to band together to defeat a proposal that they viewed as a threat to their national and collective security.

Another, similar controversy involved a proposal by the developing countries to allocate frequencies in the 14.5- to 15.35-GHz band for BSS uplinks [26]. This approach, which would have made the BSS application primary and terrestrial applications of the frequencies secondary, again interfered with radar and other defense communications used by the United States and its NATO allies, who, after considerable debate, agreed to downgrade radar applications in the 17.3- to 17.7-GHz band and to abandon radar use in the 14.5- to 15.35-GHz band over time.

Another controversy aired at the 1979 WARC was the adoption of a permanent approach to the management of the 12-GHz band in Region 2. The United States had refused to adopt the 1977 plan for the Western Hemisphere, partly because of the plan's lack of flexibility and partly because the 12-GHz band was already shared with FSS in the Western Hemisphere. Instead, in 1977 an interim *arc segmentation* plan was developed under which separate segments of the orbital arc were designated for the two services, and orbital positions were allotted on an a posteriori basis. The 1979 conference replaced this method by expanding the bandwidth available to the two services in Region 2 and allocating separate bands to each service. This method permitted both services to use the entire orbital arc serving Region 2 [27].

While the 1977 and 1979 conferences appeared to threaten a prolonged period of tension between the developing and developed countries in the management of satellite services, subsequent conferences saw some lessening of these tensions and a retreat from the doctrinaire adoption of a priori approaches to spectrum management.

The Pragmatic Phase: 1979 to 1988

In 1983, a Regional Administrative Radio Conference (RARC) met to continue the task of planning the use of the 12-GHz band for BSS in Region 2. This conference is significant because it did not continue the North-South acrimony of the late-1970s conferences, but found a way to combine a priori planning with a measure of flexibility.

Specifically, while the 1983 plan placed limits on satellite power output (particularly for DBS service) and allotted orbital positions in Region 2 to individual countries, it also permitted deviations from the plan so long as the deviations did not cause interference or, if some interference was likely, the user causing interference did so with the agreement of the affected parties.

Pragmatic results were also achieved by WARC-ORB-85 and WARC-ORB-88, which overcame political differences and respectively adopted a flexible approach to spectrum planning and enacted detailed procedures to implement it. The hybrid method that emerged from these conferences included a priori planning for new allotments in the FSS only, with improved coordination procedures for FSS allotments already made and a continuation of the first-come, first-served approach for other frequencies. Under the 1985/1988 method, even the planned portion of the FSS did not assign specific orbital slots and frequencies to individual countries, but instead guaranteed to each country at least one orbital position within a defined arc of the GSO. This approach left significant flexibility in the choice of specific orbital positions.

While we have spoken of the 1983–1988 era as a more pragmatic and less political period for the ITU's satellite spectrum allocation effort, North-South tensions remain and pragmatic results are often achieved in spite of those differences. These problems reflect the ongoing evolution of the ITU from a consensus-based organization controlled by the advanced countries to an organization in which developing countries increasingly seek ways to make their numbers count.

The GSO Under Pressure: 1988 to 1995

As the demand for positions in the GSO has grown, the ITU's effort to manage the resource through registration and voluntary coordination among users has come under considerable strain. Two problems are of particular concern: the failure of some users of the resource to complete the (admittedly frustrating) coordination process and the rise of so-called paper systems.

The first problem is typified by the 1994 launch of a Chinese satellite into an orbital slot with only one degree of separation from two adjacent satellites. The Chinese operator, APT, had not secured the agreement of the two adjacent operators to occupy the slot, and the launch violated the ITU's recommended spacing of at least two degrees between satellites. The Chinese launch occured at a time when Russia and Indonesia both claimed orbital positions within one degree of each other, raising the possibility that one of those operators would try to preempt the other with an uncoordinated launch.

The second problem is represented, most notoriously, by the Kingdom of Tonga, which has filed for orbital slots well in excess of its internal needs in order to lease them to operators for profit. The problem is also illustrated by other countries that have placed claims to orbital locations in the ITU registry for "paper systems" that are never built. The proliferation of such claims, which the ITU has no present authority to disallow, has exacerbated the scarcity of choice slots and undermined the authority of the ITU process.

A number of suggestions for coping with these problems have been made, including granting the ITU the power to reject applications that it deems unqualified and requiring countries that have registered slots with the ITU to use them within some reasonable time. (At present, countries have nine years to use a registered orbital location before its priority over other users is lost.) At a Plenipotentiary held in Kyoto in late 1994, the ITU appointed a group to study these problems and recommend solutions.

Complying With the ITU Process

Finally, since the foregoing has been somewhat theoretical and historical, we should not leave the subject of the GSO without describing in practical terms the process by which a potential user complies with the ITU's requirements for use of a GSO position. This section describes the process for both FSS and BSS applications.

If you wish to provide voice, data, or nonbroadcast video services through your own satellite, you will want to launch and operate a satellite in the FSS. How you will accomplish this depends on whether you will use a planned or unplanned FSS segment of the GSO.

If your country was a member of the ITU in 1988, it was allotted a so-called nominal FSS orbital position and a range within which that position could be adjusted. If operation within that arc and at the allotted service area and frequencies will be efficient for your planned service, you should approach your nation's regulatory agency and follow the procedures for assignment of the planned slot to your satellite. If you can operate within the parameters of the plan and are awarded the slot by your administration, you can avoid the complex coordination procedures required for use of the unplanned resource.

If no planned FSS position is available to you, then you must follow the procedures set out in Articles 11 and 13 of the ITU Radio Regulations. Of the three steps required by the Regulations (advance publication, coordination, and notification), the coordination step is the most difficult. Coordination requires that you negotiate with the operators of nearby satellites to ensure that your use of the proposed frequencies will neither cause nor receive interference. This process can take years to complete.

If your intention is to broadcast audio or video signals for direct reception by the general public, you must use an orbit and spectrum allocated to the BSS. Since all BSS orbital locations and frequencies are planned—that is, allotted to particular countries—you must be assigned a slot by the country to which it is allotted. If the country you will serve is in Region 1 or 3, this means that you must be assigned the precise orbital slot assigned to that country. If the country you will serve is in Region 2, you may have some flexibility to seek assignment of an orbital position within the range set by the orbital arc allotted to that country.

3.3 NONGEOSYNCHRONOUS ALTERNATIVES: LOW-EARTH-ORBIT SATELLITES

With all of the attention given to the management of the GSO, it is easy to forget that the first artificial Earth satellites—beginning with Sputnik in 1957—did not use the geosynchronous orbit. Instead, the orbits of the earlier military and research satellites were low and inclined at an angle to the equator. These satellites appeared to move constantly in relation to the surface of the Earth.

Historically, the great disadvantage of nongeosynchronous satellites was that they had to be tracked continuously from the ground. This required ground-based tracking stations to be placed at many locations along the satellite's track; each station would "acquire" the satellite, receive data from it, and then hand off the information to the next tracking station under the satellite's moving footprint. This approach was both expensive and impractical for communications satellite applications.

In the 1980s, however, military research devised a method of handing off data directly from one satellite to another. Now when a satellite in nongeosynchronous orbit dipped below the horizon, it could send its data seamlessly to another satellite coming over the opposite horizon. This approach eliminated the need for tracking antennas and coordination among ground stations.

The practical availability of the nongeosynchronous, low Earth orbit brought about the development of two categories of low-Earth-orbit (LEO) systems, generally called Big LEOs and Little LEOs. Big LEOs are multisatellite systems operating above 1 GHz that will provide voice and high-speed data communications as well as mobile communications to small, handheld terminals similar to those used in cellular telephone systems. Little LEOs are also multisatellite systems and will provide low-speed digital communications and position location services to portable terminals at frequencies below 400 MHz.

The 1992 WARC was the first WARC to consider the allocation of frequencies for mobile services provided from LEO satellites. WARC-92 allocated substantial bandwidth to the Little LEO systems, with the reservation that the frequencies be used only for mobile satellite services served from nongeosynchronous orbits. The conference also allocated spectrum for Big LEOs, and permitted that spectrum to be used for mobile services served by geosynchronous or nongeosynchronous satellites [28].

As the 1995 World Radiocommunications Conference approached, the U.S. mobile satellite industry urged the U.S. delegation to press for additional spectrum for mobile satellite systems. Some of those proposals, including the elimination of rules that give priority to fixed satellite systems in certain frequency bands, point to a growing divergence of interest between the U.S., which leads in the development of low-Earth-orbit and other mobile satellite systems, and other countries that remain more concerned with the availability of frequencies for fixed satellites.

Notes

[1] Some commentators have suggested, however, that emerging *spectrum-skipping* technologies may make the allocation and allotment of frequency obsolete. See G. Gilder, "Auctioning the Airways," *Forbes ASAP*, April 11, 1994, p. 9.

[2] Transmissions to and from satellites may also interfere with terrestrial microwave communications.

[3] See Chapter 2.

[4] The ITU has divided the world into a number of regions for radio regulation purposes. These are Region 1 (Africa, Europe, the Middle East, the former USSR, and Mongolia), Region 2 (Western Hemisphere), and Region 3 (Asia and the Pacific).

[5] Deviations from the ITU's allocations are subject to the noninterference obligation discussed below.

[6] The ITU recognizes that transmitting stations that are incapable of interfering with stations of other countries need not observe the ITU's regulations. See, for example, G. Codding, Jr., and A. Rutkowski, *The International Telecommunications Union in a Changing World*, Norwood, MA: Artech House, 1982, pp. 273–274.

[7] As Codding points out, a posteriori methods historically "worked to the disadvantage of those countries which have been hindered, for one reason or another, in the development of their national radio communication networks, and worked to the advantage of those countries which were foresighted enough to make exaggerated claims to frequencies." G. Codding, Jr., *The International Telecommunications Union: An Experiment in International Cooperation*, Leiden: E. J. Brill, 1952, p. 192.

[8] While global allotments generally have not been made on an a priori basis, a number of regional allotments have been made in this way. Some of these regional allotments have been made at regional conferences under ITU auspices. As we discuss below, for example, allotment of 12-GHz satellite channels was settled in 1977 for all but the Western Hemisphere, which developed its plan at a regional conference convened in 1983.

[9] Where a registration appears to interfere with an existing use, the IFRB may try to arrange a solution through discussions with the interested parties.

[10] The most crowded regions of the GSO are those serving Western Europe (1 degree west to 35 degrees west longitude), North America (135 degrees west to 87 degrees west longitude), and Eastern Europe and Russia (49 degrees east to 90 degrees east longitude).

[11] A number of hybrid satellites capable of receiving and transmitting on both C- and K-bands have also been launched.

[12] Various technical "fixes," including improvements in antenna design, can reduce the spacing requirements.

[13] A complete GSO traverses about 265,000 km, or 736 km for each degree of longitude.

[14] See *In the Matter of Licensing Space Stations in the Domestic Fixed-Satellite Service*, 93 F.C.C. 2d 1261 (1983).

[15] Yet the international community had known for some time that artificial Earth satellites would be launched. Both the United States and the Soviet Union had announced their intentions as early as the summer of 1955. R. White and H. White, Jr., *The Law and Regulation of International Space Communication*, Norwood, MA: Artech House, 1988, pp. 111–112.

[16] None of this first spectrum allocation was intended for telecommunications applications.

[17] The conference also allocated another 3,276 MHz to space services other than the communications-satellite service.

[18] The conference also adopted a coordination procedure for satellite systems, but those procedures were nonbinding and applied only to international—not domestic—communications satellite systems.

[19] The only sanction provided was for failure to respond to notification: nonresponding countries waived their right to complain about interference from the notifying country's Earth station.

[20] The 1971 WARC also allocated spectrum to other space-based uses, including Earth exploration and maritime-mobile service that became the basis for the International Maritime Satellite Organization (INMARSAT).

[21] In some cases, the satellite approach is entirely practical. The Indonesian Palapa satellite network, for example, connects hundreds of islands that would be prohibitively expensive to serve with terrestrial facilities.

[22] See note 4 above. Permanent planning for Region 2 was taken up at a regional conference held in 1983.

[23] Ten years after the plan was adopted, there was still no BSS service to Europe or Africa, although countries in those continents had a priori rights to orbital spectrum allocations for BSS.

[24] One of the odder features of the 1979 conference was the attempt of some equatorial countries to assert sovereignty over the portions of the GSO lying above their territories—a position wholly at odds with the Outer Space Treaty of 1967.

[25] A notable case was India's attempt to launch a meteorological and communications satellite. The most efficient positions were already claimed by INTELSAT and the USSR, and India was forced to use a less efficient position at lower power.

[26] As we noted earlier, the 1977 plan allocated frequencies only for the BSS downlinks.

[27] This solution still left a band of frequencies to be shared by the two services.

[28] The U.S., which was the chief proponent of the Big LEO allocation, met substantial opposition from countries that saw the systems as a threat to their control of voice services offered to customers within their borders. The Little LEO systems, with their more limited range of services, were far less controversial at WARC-92.

4 The International Telecommunications Satellite Organization

The International Telecommunications Satellite Organization, better known as INTELSAT, is headquartered in Washington, D.C. It is a multinational consortium of countries and their telecommunications providers that owns and operates 23 satellites. All the satellites are used commercially and for exclusively civilian purposes. Membership is open to any country that is a member of the ITU, but nonmembers may also use the space segment [1]. There are over 300 authorized users of the system, who may communicate over more than 27,000 Earth stations worldwide. The organization has jurisdiction over the space segment only. It does not construct, finance, or maintain the Earth stations needed to communicate with the system [2]. Approximately two-thirds of the world's international telecommunications traffic is carried by INTELSAT [3].

INTELSAT is a product of its time. The idea for a global satellite system with universal coverage originated in the optimistic years that followed World War II, when international cooperation seemed both possible and beneficial. Its growth and success are not only a tribute to the mentality of that era, but also the result of a long history of necessary cooperation among international telecommunications operators. Any transborder communication necessarily involves two operators who are forced to interconnect and cooperate in order to complete the transmission. For that reason, telecommunications operators around the globe began working together early in the history of communications by jointly owning facilities, setting rates, promulgating standards, and routing traffic. INTELSAT has greatly benefited from this history of cooperation. It is reflected in the decision-making process within the organization. INTELSAT members are exhorted to reach decisions by consensus, and without such consensus, a supermajority is generally needed to make most decisions.

INTELSAT was created as a global satellite communications monopoly in 1963. As an intergovernmental organization, it was then and is now immune from antitrust prosecution. Its mission reflects a mixture of idealism and pragmatism. Its goal was to implement a single global system that would provide public communications to all areas of the world, including those that are geographically remote or difficult to access due to topography. An additional goal was to contribute to world peace and understanding by promoting the use of "the most advanced technology available" and by using "the most efficient and economic facilities possible consistent with the best and more equitable use of the radio frequency spectrum" [4]. It contemplated participation by all nations of the planet, rich or poor. However, that last goal was not achieved until recently, when the collapse of the former Soviet Union resulted in the lifting of the ban on organization membership imposed by the Soviet Union on Eastern Bloc countries. INTELSAT is currently still the only satellite system with global universal service and nondiscrimination obligations. Recently, its status as a monopoly and the principles of one country, one vote have been increasingly called into question by separate satellite systems, telecommunications administrations, some governments, and the organization itself.

This chapter discusses the creation and function of INTELSAT in its historical context, its current organization and structure, its workings, and the prospects for its future survival in an era of declining monopolies. Particular attention is paid to the issue of eligibility for membership as well as to the ongoing debate concerning direct access and membership by entities other than each country's signatory.

4.1 A HISTORICAL PERSPECTIVE

The invention of the electric telegraph by Samuel Morse in 1836 marked the beginning of the telecommunications revolution. The technological changes fueled by the Industrial Revolution soon underscored the need for a cheap, fast, and reliable technology to provide telecommunications throughout the globe. Although undersea cables were introduced in the middle of the nineteenth century, their technology had limited coverage and capacity. At that time, they were also comparatively more expensive and subject to frequent breaks that were disruptive and difficult to repair. The direct predecessor of satellite communications, introduced in the 1920s, was shortwave radio transmissions that traveled over long distances by bouncing off the ionosphere. Although this means was theoretically capable of relaying voice messages across continents, it was very unreliable, and the voice quality was bad. Ionospheric radio transmissions did, however, constitute somewhat of an improvement over then-existing cable systems in terms of coverage and cost [5]. The first workable proposal for a system of geostationary satellites that would provide global coverage was not introduced until shortly after World War II [6]. Arthur C. Clarke, a British physicist and science fiction writer, observed in 1945 that if an object is placed in a certain orbit around the Earth, it will have an orbital frequency of 24 hours, will move in consonance with the terrestrial globe, and will therefore be constantly over the same point. This orbit is known as the geostationary orbit, or the Clarke orbit [7].

INTELSAT was essentially a U.S. creation. On the other hand, the enormous costs associated with telecommunications facilities and operations (such as the installation of undersea cables and their maintenance and repair) and the international nature of communications have fomented from the beginning an unusual climate of technical and commercial cooperation between governments and international telecommunications operators. It is this history of cooperation between telecommunications operators and authorities that is largely responsible for the relatively painless development and commercial success of INTELSAT.

The first experimental communications satellites were launched in the fifties. By 1960, the United States had realized the potential of this new technology and was determined to encourage its development. President Eisenhower was the first to accord full support to development of commercial satellite technology for telecommunications use. In a 1960 statement, President Eisenhower said: "[t]he Government should aggressively encourage private enterprise in the establishment and operation of satellite relays for revenue producing services" [8]. Soon afterward, President Kennedy called for the creation of a satellite consortium partly owned by foreign nations that would guarantee nondiscriminatory access to the system to all nations of the world. In response, the United Nations passed a resolution declaring that "communications by means of satellite should be available to the nations of the world as soon as practical, on a global and nondiscriminatory basis" [9]. This statement lay the groundwork for the creation of INTELSAT as it is known today.

Because the United States was the only country with viable communications satellite technology, it soon took the lead in this area. Initially, the organization was to be a U.S. corporation, fully owned and controlled by U.S. interests. Three proposals were introduced before the U.S. Congress in 1962 to create the new satellite organization. The first proposal was for a communications corporation composed exclusively of U.S. common carriers [10]. Understandably, this proposal was supported by the carriers themselves. The second proposal, introduced by the Kennedy administration, represented a compromise between the goal of establishing an international organization and the perceived need to ensure U.S. leadership in the project. Under that proposal, ownership would be divided between publicly traded stock and stock held exclusively by U.S. common carriers. The proposal was finally adopted as the Comsat Act of 1962 [11]. The third proposal was for a satellite system entirely owned by the government, with no private enterprise participation [12].

The European reaction to those U.S. efforts was to create the European Conference of Post and Telecommunications Administrations (better known by its French initials, CEPT) in 1962, later replaced by the European Conference on

Satellite Communications. This group took a unified position in all telecommunications matters, often in opposition to the views of the United States.

At the same time, international efforts to launch a global satellite network were underway under the auspices of the ITU. This idea was only reluctantly accepted by the United States, fearful of losing its avowed edge in the satellite area. The intent of the United States had been to construct a commercial global satellite system based on a series of bilateral agreements (much as the airline industry today). The unified approach taken by the Europeans in opposition to the United States made bilateral bargaining increasingly difficult [13]. Reluctantly, the United States accepted the idea of an international consortium. However, it insisted that the Soviet Union and the developing countries be excluded from those negotiations [14].

Finally, in 1964, the United States and the European Conference on Satellite Communications sat at the bargaining table and concluded the two agreements that were the predecessors of the documents that form the basis of INTELSAT today: the Intergovernmental Interim Agreement [15], which was the predecessor of the INTELSAT Agreement [16], and the Special Agreement, which was at the origin of today's Operating Agreement [17]. The parties also agreed that Comsat Corporation would temporarily manage INTELSAT. At that point, the Soviet Union was offered membership, but for reasons that are not entirely clear, it declined [18].

Because of the U.S. lead in satellite technology, it held then and holds now the largest interest in the organization. Although developing countries were eventually admitted into membership, a cap was placed on their ownership. At the same time, the U.S. dominance of the organization and the fact that most European countries had invested heavily in cables remained a source of friction with the Europeans, as did procurement issues. The Intergovernmental Interim Agreement provided for ownership in the system based on usage and capital contribution to the organization for design, maintenance, and operation of the system, as well as for the operating costs of the organization. It was initially signed by 11 nations: Australia, Canada, Denmark, France, Italy, Japan, the Netherlands, Spain, the United Kingdom, the United States, and the Vatican City [19]. At the insistence of the European nations, who wanted to leave open the possibility of constructing their own satellite system later, the INTELSAT Agreement contains a clause permitting the organization to authorize other satellite systems separate from INTELSAT (these systems are known as *separate satellite systems*). INTELSAT will authorize the construction of a separate satellite system only after the proponent has demonstrated that no technical or economic harm will be done to the organization by the new entrant. Separate satellite systems are dealt with in more detail in Section 4.5 and in Chapter 5. The Special Agreement was signed by the telecommunications operators of every member country. Under these interim agreements, INTELSAT was entirely managed by Comsat without international participation for a fee of

$150,000 a year. The final agreements were signed in 1971 and went into effect in 1973. As mentioned above, the Interim Agreement became the INTELSAT Agreement [20], and the Special Agreement became the Operating Agreement [21]. These new agreements curtailed the role of Comsat and introduced international management and satellite procurement into the organization.

The first INTELSAT satellite, Early Bird (INTELSAT series I), was launched in 1965. With 480 telephone channels, Early Bird had roughly twice as much capacity as the largest existing undersea cable [22]. This satellite was followed by INTELSAT series II, III, IV, V, VI, VII, K (an all Ku-band satellite used for broadcasting and the first direct-to-home satellite broadcast service), and VIII, to be launched beginning January 1996. INTELSAT now offers a variety of telecommunications services, including domestic and international telephone service [23], analog and digital service, private network applications, video service with associated audio, service on a long-term lease basis or for occasional use, cable restoration services, and direct-to-home satellite broadcasting service. INTELSAT's clients include the UN, which leases capacity on Atlantic and Indian Ocean region birds. INTELSAT will also implement ASETA's regional communications plan (Association of Telecommunications' State Enterprises of the Sub-Andean Agreement, Bolivia, Colombia, Ecuador, Peru, and Venezuela).

4.2 THE INTELSAT AGREEMENTS

As mentioned in Section 4.1, the INTELSAT Agreement and the Operating Agreement form the basis of INTELSAT. They contain provisions relating to the governance of the organization, accession to the treaty, who may become a member, and relationships between signatories and parties, parties and parties, signatories and signatories, and signatories and parties and the organization. The organization has four governing bodies: the Assembly of Parties, the Board of Governors, the Assembly of Signatories, and the Executive. The cost of participating in the meetings and assemblies are borne by the participants and are not part of the overall budget of the organization.

4.2.1 The Assembly of Parties

The Assembly of Parties is one of INTELSAT's two democratic organs. It is composed of representatives of each country that adheres to the INTELSAT Agreement, regardless of size or investment share. Each party represents the national interest of its country as a sovereign state and has the right to cast one vote on its behalf. The Assembly of Parties meets once every two years to consider documents, resolutions, opinions, or recommendations submitted to it by the Board of Governors or the Meeting of Signatories. Extraordinary meetings may be called by the Board of Governors or by any party with the concurrence of at least one-third of the parties. The Assembly of Parties is responsible for discussing and deciding the general policy and long-term objectives of the organization and for ensuring that any such plans do not conflict with any international treaty consistent with the INTELSAT Agreement and adhered to by at least two-thirds of the parties. It is also the organ responsible for conducting the organization's foreign relations policy. As such, it decides whether to establish formal relationships between INTELSAT and states, whether parties or not, and with other international organizations, and whether INTELSAT will participate in a given conference pursuant to an official invitation. Although any party may submit a proposal to amend the INTELSAT Agreement, only the Assembly of Parties may modify the agreement, and this by a two-thirds majority vote. Any such modification must also be ratified by two-thirds of the states that adhere to the INTELSAT Agreement and that hold a total of at least two-thirds of the total investment shares in the organization. A simple majority is needed to make most decisions, except when the matter is one of substance or of particular importance. In that case, a two-thirds majority is needed. Any consultation under Article XIV of the INTELSAT Agreement [24] must receive the approval of the Assembly of Parties.

The Assembly of Parties considers and decides complaints submitted by parties about other parties or about signatories. If the dispute is not resolved to the satisfaction of the complainant, it may appeal to an arbitration tribunal composed of three members selected by the Assembly of Parties [25]. Any party may withdraw voluntarily from INTELSAT at any time, with or without a reason. Under some circumstances, the Assembly of Parties may terminate a party's membership in the organization upon notification or on its own initiative [27]. However, no party has ever been expelled. In addition, because the main purpose of INTELSAT is to ensure the availability of public telecommunications services [27], those services are given priority in assigning capacity. The Assembly of Parties must be consulted in order to authorize the use of the space segment for specialized telecommunications services such as broadcast satellite television, radio navigation services, or meteorological services.

Finally, the Assembly of Parties adopts and modifies the organizational structure of the executive and appoints the director general. In spite of the fact that every country is represented and has equal voting power in this body, since the inception of INTELSAT, the director general has always been an American, generally from Comsat.

4.2.2 The Meeting of Signatories

The Meeting of Signatories is convened once a year. It is composed of a representative of each signatory to the Operating Agreement, that is, of a representative from each telecommunications operator of each member country. Each signatory has one vote, regardless of investment share. The Meeting of Signatories adopts its own rules of procedure and elects its own chairman. Decisions are made by simple majority. By far its most important function is the determination, every year, of the minimum investment share that entitles a signatory to have a governor on that year's Board of Governors rather than having to share one with a given group of signatories. In addition, because its members are INTELSAT investors most directly affected by the economic performance of the organization, and because as telecommunications operators they have the technical expertise, the Meeting of Signatories has mostly financial and technical responsibilities. It also participates in INTELSAT management through its recommendations to the Board of Governors concerning the structure of the organization's executive. The Meeting of Signatories makes general rules submitted to the Board of Governors concerning the approval of new Earth stations authorized to use the INTELSAT space segment, allotment of space segment capacity, and the rates to be charged to users of the space segment. It also takes decisions concerning the priority to be given a domestic telecommunications project considered under Section (b)(ii) of Article III (see Section 4.4). The Meeting of Signatories also presents the Board of Governors with its views regarding financial issues, such as the monetary implications of proposed future programs, the annual reports or annual financial statements of the organization, and the capital ceiling provided for in Article 5 of the Operating Agreement. (For more on the capital ceiling, see Section 4.3) [28]. It also reports regarding the implementation of INTELSAT's general policies, activities, and long-term programs.

Any signatory may withdraw from INTELSAT at any time, but the Meeting of Signatories may terminate a signatory only under certain circumstances, such as failure to meet its obligations under the agreements. The Meeting of Signatories also presents its views on complaints submitted by signatories or nonsignatory users of the system regarding other users, signatories, or the organization itself.

4.2.3 The Board of Governors

The Board of Governors is the true executive and policy organ of INTELSAT. Currently, 28 governors represent 111 signatories [29]. This is the least democratic of all the INTELSAT organs, since only signatories with a certain investment share are entitled to be represented by a governor. The board adopts its own rules and elects its own chairman. The Board of Governors is the governing body of the organization. It is responsible for planning, designing, constructing, and operating the space segment and for any other activities that INTELSAT may carry out under the agreements. Any procurement project with a value exceeding US$500,000 must be approved by the board. In that case, the board must also give its approval before any such contract is awarded to a particular bidder. It also approves the criteria adopted by the Meeting of Signatories concerning Earth station approval and technical standards. Its responsibilities include deciding the terms and conditions of space segment allotment. In addition, it decides the number, status, and employment conditions of the senior executives and appoints the director general subject to confirmation by the Assembly of Parties.

The number of governors is restricted by the INTELSAT Agreement, which admonishes the Meeting of Signatories to keep it as close to 20 as possible. In order to ensure that this goal is accomplished, the agreement details a complicated method of determining the minimum investment share that will entitle a signatory to a governor each year. To determine the minimum investment share, the Meeting of Signatories must take into account the number of governors elected to the last Board of Governors meeting. The organization prepares a list of signatories ranked by investment share in decreasing order. If the number of governors at the last board meeting was less than 20, the minimum investment share is that held by the signatory whose share was immediately below that of the last signatory to receive a governor at the last board meeting. For example, if at the last board meeting Singapore was the last signatory entitled to a governor, with a rank of 15th on the list and investment share of 1.666% this year, the minimum investment share will be 1.59, or that held by the Netherlands, the signatory who is next on this year's list.

If the governors at the last meeting numbered 20, 21, or 22, the minimum investment share is that held by the signatory who holds the same position as the last signatory to qualify for a governor at the last board meeting. If, however, the number of governors exceeds 22, the minimum investment share is that held by the last signatory ranked above and whose investment exceeded that of the last signatory to obtain a governor in the previous year's board meeting. Article IX contains an escape clause allowing the Meeting of Signatories to employ other methods of computing the minimum investment share in case the method contained in the article fails to ensure that the number of governors remains close to 20.

Signatories whose investment share does not qualify them to have a governor do not remain completely unrepresented. Each group of two or more signatories that do not have a governor may be represented by one governor for the entire group if its combined investment share equals or exceeds the minimum investment share. For example, in 1993 Argentina had a share of 1.788719%, Chile had a share of 0.819815%, and Uruguay had a share of 0.050000%. Individually, each country had less than the minimum share needed to have a governor. However, these three countries combined had a share of 2.658594%, enough to earn them a governor as a group. Finally, for signatories otherwise unable to obtain representation, one governor will be named upon request by the group to the INTELSAT executive for every group of at least five signatories that belong to one of the regions defined by the ITU's Plenipotentiary Conference of 1965, held at Montreux, Switzerland. However, that number may not exceed two governors per region or a total of five. If the number of requesting groups exceeds those limits, the Meeting of Signatories will determine which groups will be represented. Priority is given to the groups with the largest combined investment shares. On the other hand, governors appointed under this system are not counted in determining the total number of governors in calculating the number of governors for the purpose of establishing the minimum investment share pursuant to the list. Representation on the Board of Governors continues for the year even if the investment share of a signatory or a group of signatories falls below the minimum level, but it ceases if the drop is caused by the withdrawal of a signatory from the organization or from the group.

Unlike the case in the other governing bodies, voting in the Board of Governors is weighed with relation to the investment share held by the signatories represented by a given governor in the INTELSAT space segment used for international public telecommunications services, domestic telecommunications services between areas separated by other countries, or to remote areas with insufficient landline facilities [30]. However, no governor may hold a voting power of more than 40%. Thus, for example, the United States, which has its own governor and an investment share of 20.149553%, has a voting power of 20.149553%, while the Argentina-Chile-Uruguay group, which has only one governor for three signatories, has a voting power of 2.658594%. Ideally, the Board of Governors should make decisions by consensus; but if that proves impossible, a decision on substantive questions is reached with the agreement of at least three governors having at least two-thirds of the vote or of all but three of the governors, regardless of voting power. For procedural matters, a simple majority of governors without regard to investment share suffices.

4.2.4 The Executive

The director general is INTELSAT's chief executive, and answers directly to the Board of Governors. The current director general is Mr. Irving Goldstein of the United States, formerly with Comsat. The organization is supported by a large multinational staff composed mainly of engineers. The INTELSAT executive manages the day-to-day operations of the organization.

4.3 INTELSAT MEMBERSHIP AND FUNCTION

4.3.1 Membership

As noted earlier, any country that is a member of the ITU is eligible to become an INTELSAT member. As of January 1995, 135 nations were members of INTELSAT [31]. Because the organization is at once an intergovernmental entity and a commercial enterprise, each country's government is represented in the Assembly of Parties and its telecommunications operator in the Meeting of Signatories. INTELSAT functions as a cooperative. Each signatory must make a capital contribution of at least 0.05% of the valuation of the organization upon joining [32].

Afterwards, for as long as the signatory continues to be a member, it must make regular contributions based on its use of the system as needed to meet INTELSAT's capital requirements to build and maintain the space segment and other INTELSAT property. Article 5 of the Operating Agreement limits the sum of the net capital contribution of signatories and the capital commitments of the organization to a sum called the capital ceiling. The initial capital ceiling was $500 million, but the Board of Governors in conjunction with the Meeting of Signatories may raise this ceiling as appropriate. In 1994, the capital ceiling was $4 billion.

Each signatory's investment share is calculated on March 1 of each year based on its actual percentage of satellite use during the previous 180 days relative to the use by all signatories. Since INTELSAT is a cooperative, the sum of all the outstanding shares must always equal 100%. The minimum investment share is 0.05%, regardless of use. In order to calculate investment shares, the percentage of the total number of minimum investment shares is subtracted from 100%, and that figure is apportioned among the remaining signatories based on use. For example, in 1993, 32 of the 133 members of INTELSAT held the minimum investment share. Thus, 1.6% of the organization was in their hands. The remaining 98.4% share was apportioned among the other 101 signatories according to their use of the space segment. France Telecom was found to have a 3.943627% share, British Telecom had a 10.989744% share, and Comsat Corporation, the U.S. signatory, had a 20.149553% share. When a new signatory accedes to the organization, the investment shares of the existing signatories holding more than a 0.05% share are reduced accordingly to ensure that the organization stays within the capital ceiling. Similarly, when a signatory withdraws from INTELSAT, its share is apportioned among the remaining signatories holding over 0.05%, who are required to buy a portion based on their investment share. A signatory may request a larger or smaller share than its use of the system warrants under Article 6 of the Operating Agreement. As explained in Section 4.3, a greater investment share confers upon a signatory greater voting weight in the Board of Governors, and may make the difference between having and not having a governor to represent its interests. In addition, as will be seen below, a greater investment share entitles a signatory to a larger portion of the organization's revenue distributions. Since the total of investment shares must always be 100%, a request for a change in share size can only be accommodated to the extent other signatories also wish to adjust their shares. A request for a share that is higher than warranted by a signatory's use will always be granted if possible. A change to a lower share will only be acceptable if the resulting share is no less than the minimum investment share and not less than the signatory's share of use on the last day of February (the day before the new investment shares are set). When a request for a change in investment share cannot be fully accommodated, the organization may partially grant it to the extent it is consistent with Article 6 of the Operating Agreement. If, for example, in a given year, shares totaling 4% are made available by signatories desiring a lower investment share, but the demand for an increase of investment share relative to use totals only 1%, 3% in shares would not be absorbed, and the signatories desiring to reduce their share would only be partially able to do so. Conversely, if the demand for an increased share totaled 3%, but only 2% were made available by signatories desiring to reduce their share, the signatories who want to increase their participation would only be able to increase it accordingly. Historically, however, with only one exception [33], requests for lesser shares have always been offset by requests for greater shares. Each share has a monetary value denominated in U.S. dollars. Naturally, the value of each share is its percentage of the net worth of INTELSAT (or *valuation*) on any given day. Thus, the value of a signatory's investment share is easily ascertainable by multiplying the signatory's share by the INTELSAT valuation on the day of the transaction.

Each quarter, INTELSAT calculates the capital it needs to meet its expenses for the upcoming quarter and bills each signatory for a portion of that sum based on its investment share. In addition, signatories and other users are billed for their use of the space segment. If a signatory fails to pay its capital contribution within three months of the due date of the bill, it is considered to be in default under the agreements and its rights are suspended until payment is made. At the end of each quarter, INTELSAT subtracts its operating costs from its revenues and distributes the remainder between its signatories. Under Article 8 of the Operating Agreement, the Board of Governors sets annually a target rate of compensation for the use of the signatories' capital for the year, taking into account the risk of investment in INTELSAT and the price of money in the world markets. Naturally, the rate of compensation is a factor of a signatory's investment share. The greater the investment share, the greater the compensation. In 1994, the Board of Governors set the rate of compensation at 20%. Any sums remaining at the end of the year are retained as excess compensation and used as a reserve account to offset launch failures and other catastrophic events.

4.3.2 INTELSAT Access and Use

In order to connect into the system, a user must obtain access from the signatory. Here, a distinction between the United States and the rest of the world is in order. The United States created Comsat Corporation in 1962 as the U.S. signatory to INTELSAT and an intermediary between the users of INTELSAT capacity and the organization. No U.S. carrier has ever been authorized to access the INTELSAT space segment directly. Comsat charges every carrier an access charge (or *markup*) for its role as gatekeeper to the system. Comsat itself is prohibited by law from becoming a carrier. While some U.S. international carriers are frustrated by the existence of this intermediary, the FCC imposes on Comsat the obligation to treat all its customers equally, without discrimination. U.S. carriers therefore enjoy a level playing field with respect to each other. In other countries where competition has been introduced, the signatory is often the user's major competitor. This situation creates an uncomfortable tension between the established operator and the market newcomers.

When INTELSAT was created in 1964, telecommunications operators were monopolies in every country, including the United States. It made sense, therefore, to have a single signatory per country. However, nowhere in the INTELSAT Agreements is it specified that there should be one signatory per country. In the United States, the monopoly of the Bell System did not begin to disappear until the 1980s. But even the breakup of the long-distance and international monopoly in the United States did not have an effect on INTELSAT membership qualifications. As mentioned above, the Communications Act of 1962 created Comsat as a corporation whose sole purpose is to act as the intermediary between INTELSAT and the U.S. international carriers, and gave it a perennial monopoly. The United States is the only country to have such an entity. Therefore, the introduction of competition in the United States had no effect on INTELSAT membership or the qualifications required to become a signatory.

On the other hand, the Agreement and the Operating Agreement do not limit the number of signatories to one per country. In the 1990s, a number of countries have introduced competition in their long-distance and international markets. Some of those countries—including Chile, Argentina and the United Kingdom—have gone one step further and have liberalized access to the INTELSAT space segment by making it lawful for any Earth station licensee to access the system directly, without the intermediary of the former monopoly national telecommunications operator. Operators who are authorized to have direct access to INTELSAT but are not signatories are known as *designated clients*. There are four levels of direct access. In each case, the signatory must notify INTELSAT of the level of access it will permit the designated client to have.

The first, and lower, level of access is operational and technical access. Under this mode of access, a user may attend the annual Global Traffic Meeting and the Global Operational Representatives Conference, two important annual meetings organized by INTELSAT, where representatives from telecommunications operators mingle and discuss business plans and future relationships. At the Global Traffic Meeting, INTELSAT users provide the organization with their nonbinding traffic forecasts for the upcoming years. INTELSAT uses this information in planning future satellite design and responses to other customer needs. Operators with this level of access may hold meetings with INTELSAT staff to discuss technical issues and request carrier assignments. Once an operator has received this level of access, it no longer has to request technical assignments through the signatory.

The second level of access is commercial and operational access. Under this mode of access, operators may discuss directly with INTELSAT staff commercial and service issues such as capacity availability, INTELSAT tariffs, service terms and conditions, and the policies concerning reservation of capacity.

The third level of access is contractual access. Operators with this level of access may directly order all INTELSAT services and undertake long-term commitments for multiple-use leases (MUL) or specific services. They may also submit requests for approval of their Earth stations. INTELSAT will contact this client directly concerning all service aspects, and it will bill the client directly for satellite utilization charges. However, unless the signatory has specifically transferred its responsibility for payment to the client, the signatory will remain ultimately responsible for payment of those charges should the client fail to meet its obligations. When a signatory seeks to transfer this responsibility to a client, INTELSAT will carefully examine the client's financial status and creditworthiness before agreeing to deal with it directly, and may impose any conditions it deems appropriate to guarantee timely payment. The client and INTELSAT will then sign a Service Agreement detailing the terms and conditions of their relationship. Signatories may also notify INTELSAT that they will not be responsible for any damage caused to the space segment by improper use of an Earth station. The party must then accept the ultimate responsibility for damages if the client fails to meet its obligations. If the party does not accept this liability, the signatory remains responsible.

The fourth and highest level of access is investing entity status. An operator with this level of access is know as a *nonsignatory investor*. Its rights and obligations are comparable to those of signatories, as specified in the nonsignatory investor agreement it signs with INTELSAT. The investing entity's initial investment share is a part of the total utilization of the country unless it is assigned on March 1, at the time all the other investment shares are calculated. In any event, the combined investment interests of the signatory and the nonsignatory investor may not be less than the 0.05% minimum investment share. The precise status of nonsignatory investors remains somewhat unclear. The investing entity agreement states that the INTELSAT Agreement and the Operating Agreement should be read as applying to these operators "as if the word 'Signatory' were substituted with the word 'non-Signatory investor.'" However, the guidelines that apply to these operators clearly confer upon them less than equal status. Nonsignatory investors may increase or reduce their investment share in proportion to their use of the system, but they can only do so with the approval of the signatory. Their requests for share adjustments will only be considered after the requests of all the signatories have been satisfied. If the nonsignatory investor's share increases and the operator does not pay for this increase, the signatory remains ultimately liable. Moreover and more importantly, a nonsignatory investor remains so at the sufferance of the signatory. The signatory may decide to revoke an operator's status as a nonsignatory investor and absorb its investment share. In that instance, INTELSAT will cancel the nonsignatory's participation upon notification.

The apparent unfairness of the nonsignatory investor regime, whereby an operator assumes all the obligations of a signatory but receives only a portion of the rights, has prompted certain countries to urge INTELSAT to accept multiple signatories per country. The United Kingdom and Chile have been the chief proponents of this approach. The Assembly of Parties met in Argentina in October of 1994 and rejected a proposal submitted by the United Kingdom to permit multiple signatories and multiple governors per country. Instead, the Assembly of Parties announced its support of the principle of multiple signatories on a voluntary basis. Under this model, each country would be free to decide whether it will allow multiple signatories. There will be only one vote per country at the Assembly of Signatories, and there can be no more than one governor per country. This proposal falls short of the reforms needed to guarantee fairness and equality among signatories. Although it goes one step further than the nonsignatory investor concept in equalizing the rights and obligations of former monopolies and their new competitors, it still preserves the second-class status of the newcomers. Proponents of the new model argue that far from guaranteeing a more favorable position to the former monopoly than the status quo, it introduces the possibility of the newcomer becoming the signatory with the vote and the governor. Defenders of this argument tend to ignore the fact that in an organization largely built on consensus, as is INTELSAT, the long-standing personal relationships established between the party and the signatory make intrusion by a new player unlikely to succeed at best. The Assembly of Parties asked INTELSAT to form a working group to make recommendations on proposed changes to the structure of the organization, including implementation of the multiple-signatory concept.

In selecting an INTELSAT service, carriers must take into account the application and the duration of the service needed. Capacity is available on both a short-term and a long-term basis. INTELSAT offers discounts to customers entering into long-term commitments. Users may upgrade, but not downgrade, their service requests. Carriers may lease a service or obtain channel or carrier-based service. Leased services permit customers to design networks with their own equipment at a level of quality they find acceptable and within their budget. As long as the service poses no technical interference to other offerings, INTELSAT will not object. A channel or carrier-based service conforms to certain quality-of-service standards guaranteed by INTELSAT at the receiving Earth station. Carriers purchasing this service must use INTELSAT-approved Earth stations and equipment. There are two ways of reserving leased service: by means of a first-refusal reservation (FRR) or by a guaranteed reservation (GR). An FRR may be submitted up to three years in advance of the anticipated need for the capacity, or up to five years if the capacity is not yet operational. To make a reservation, a carrier must pay a reservation fee, which is forfeited if service is not implemented or the reservation is not upgraded to the higher GR level before the start date of the service. If the carrier upgrades to a GR and begins using the capacity, the reservation fee is applied against the lease charges. A GR is a firm commitment to use the capacity and therefore does not require a reservation fee. It may not be submitted until one year prior to the beginning of the service, or three years if the capacity is not yet operational. The customer assumes full responsibility for payment of the lease for the full term. In order to apply for a GR for international service, a carrier must convince its foreign correspondent to submit a matching order. No matching order is necessary when requesting an FRR. Thus, FRRs are used by carriers who are unsure of their real traffic requirements but want to preserve the option of using a certain capacity if their future needs justify it.

When a customer desires to acquire capacity that is the subject of a FRR, and its service request cannot be otherwise accommodated, it may ask INTELSAT to inform the holder of the FRR that another party is interested in that capacity. The acquiring customer must be willing to assume a GR obligation with respect to that capacity. This process is known as a *challenge*. INTELSAT will keep the identity of the challenging party confidential. The challenged party then has three options. It may upgrade its lease to a GR of equal or greater value to INTELSAT than the challenger's request, thereby defeating the challenge; request that its FRR be transferred to other available capacity; or lose its FRR to the challenger's GR and receive a refund of the reservation fee. If the challenged carrier does not respond to the challenge, the GR requested by the challenger will automatically be effected.

Channel and carrier services operate differently. There are no reservation fees, and no FRR interests. Reservations are made on a first-come, first-served basis. Channel and carrier capacity reservations may be made at the annual Global Traffic Meeting.

Within each country, there is at least one operations representative responsible for coordination of Earth stations who is the point of contact between INTELSAT and the users in case of problems or emergencies. The operational representatives approve the operational plans and the proposed Earth stations in their territory to ensure their compliance with the INTELSAT Earth Station Standards (IESS) regarding performance characteristics of the stations and the Satellite Systems Operations Guide (SSOG).

4.4 THE INTELSAT NONDISCRIMINATION AND UNIVERSAL SERVICE OBLIGATIONS

As mentioned in Section 4.1, INTELSAT is the only satellite system with nondiscrimination and universal service obligations. Although the obligations are stated in the INTELSAT Agreement, they are never defined. Article III(a) declares that the organization "shall have as its prime objective the provision, on a commercial basis, of the space segment required for international public telecommunications services of high quality and reliability on a non-discriminatory basis to all areas of the world." In practical terms, universal service is considered to encompass establishing global coverage for the system, giving priority to basic international telephony and lifeline services and reasonable rates for service. There are three tiers of service priority. The agreement specifically directs the organization to give the highest priority to international service. The aim of INTELSAT is not to compete with each country's domestic network, but to link geographically remote areas as part of its mission to contribute to world peace and understanding. Therefore, the provision of international communications is its main mission. However, to achieve the aim of contributing to global peace and understanding, certain domestic communications are given the same priority as international transmissions. Thus, an exception is made for domestic public telecommunications projects between two areas of the same country that are separated by the territory of another country or by a natural barrier such as an ocean. A second exception is made for geographically remote areas not served by land-based systems for technical reasons due to insurmountable topographical barriers such as high mountains, marshes, or volcanoes. However, the domestic projects on contiguous national territory must obtain the prior approval of the Board of Governors to receive the same priority as international services.

The second level of priority is for domestic basic telephony. The organization need only accommodate requests for this service to the extent international service is not impaired. Finally, the system may provide capacity for specialized telecommunications services as long as service in the other two categories is not compromised.

The INTELSAT nondiscrimination obligation is also never defined in the agreements. It is, however, embodied in the equal representation in the Assembly of Parties and the Meeting of Signatories and in the INTELSAT tariff structure. Tariffs for INTELSAT services are the same regardless of the parties involved. In setting the price of a particular service, the organization may only consider average costs, not the cost of establishing and maintaining a particular route. Article V of the Intergovernmental Agreement prohibits price discrimination thus: "The rates of space segment utilization charge for each type of utilization shall be the same for all applicants for space segment capacity for that type of utilization." Tariff nondiscrimination has resulted in a subsidy from wealthier nations to developing areas. In large industrialized nations, the high volume of traffic justifies building more and better facilities that could theoretically be priced at their marginal cost and still be profitable. On the other hand, developing nations typically have routes that are used infrequently (known as *thin routes*), and service on these routes, if priced in a way related to actual cost, would be considerably more expensive than for their industrialized-nation counterparts. Some authors have argued that this subsidy system is inefficient and that rather than subsidizing telecommunications development through average cost-based tariffs, INTELSAT should increase its already existing direct subsidy program for developing nations [34]. This is, however, a politically sensitive issue, since some parties oppose direct subsidies. Tariff-based indirect subsidies are a less visible and more discreet form of aid. They are also typical for monopolies with universal service obligations such as telephone companies and other utilities and are thus easier to accept and understand.

4.5 INTELSAT ARTICLE XIV AND SEPARATE SYSTEMS

INTELSAT was conceived initially as a satellite monopoly. However, as explained above, at its inception, tension erupted between the European group (CEPT) and the United States, which was then and is now the largest shareholder of INTELSAT. The Europeans wanted to preserve the option of constructing their own regional international satellite system in the future. For this reason, they insisted that the INTELSAT Agreement include an escape clause permitting the establishment of international satellite systems other than INTELSAT. Non-INTELSAT international satellite systems are known as separate systems. The escape clause is contained in Article XIV of the agreement. Article XIV requires any party that will use non-INTELSAT satellite facilities for its communications needs to demonstrate first to the Board of Governors that the facilities will not cause technical harm to the INTELSAT system. The technical compatibility requirement is eminently reasonable, and it applies to international, domestic, and special services alike. The tension described above is reflected in paragraph (d) of Article XIV, which requires parties or signatories that intend to build and operate a satellite system separate from INTELSAT for international communications to first demonstrate that the competing system will not cause significant economic harm to the system. Thus, anyone trying to build a separate international system must first obtain the approval of the Board of Governors and the Assembly of Parties by demonstrating that the new system will cause no technical damage or significant economic harm to INTELSAT.

The justification for the economic harm requirement is fairly simple. INTELSAT is the only satellite system with universal service and nondiscrimination obligations. This means that it must serve every country, regardless of cost, and that it must maintain facilities on thin routes. It also means that the system may not take into account the cost of establishing and maintaining a route in pricing its service, but must distribute those costs evenly over the entirety of its routes. The price of a particular service is generally obtained by dividing total forecast costs by total forecast traffic without regard to the traffic density on a particular route [35]. The economic harm requirement is a way of protecting the system from *cream-skimming*, a practice consisting of providing separate facilities at a lower cost on high-traffic (or *thick*) routes. The incentive for cream-skimming is obvious. It is significantly cheaper on a per-minute basis to serve a thick route than a thin one. These routes are highly profitable and customers are easy to find. Because separate systems are not saddled with the cost of serving thin routes, it is easy for them to operate at a lower cost than INTELSAT on profitable routes and to pass the savings on to their customers. Other detractors of separate systems argue that INTELSAT is a natural monopoly, and that permitting separate systems would result in unnecessary and wasteful duplication of facilities and in a diversion of resources that could be better used in enhancing the quality of current telecommunications routes and links. Separate systems also endanger the INTELSAT internal subsidy system embodied in the INTELSAT tariffs as described in Section 4.4 above.

On the other hand, the introduction of competition to INTELSAT has undeniable advantages for consumers. It is a well-documented economic fact that companies respond more swiftly and flexibly to customer needs in a competitive environment. In the case of INTELSAT, this competition has resulted in the development of VSAT service to provide direct, focused communications to end users at their place of business and bypass the public switched network and the land-based national network otherwise used to relay the information from the INTELSAT Earth station to the end user. The threat of substantial competition has also fueled the rapid development and launch of INTELSAT-K, the first all Ku-band satellite to be used to provide DBS service. INTELSAT-K's principal competitor was Alpha Lyracom's PanAmSat 3 (PAS-3) satellite, which was lost due to a launch failure in December of 1994. Competition will also encourage INTELSAT to realize necessary efficiencies to stay competitive and to price its services as close to cost as possible to minimize the incentives for cream-skimming.

The first separate system to receive approval was EUTELSAT in 1979, which was soon followed by PALAPA and ARABSAT. Several U.S. systems have also been consulted. The United States has been largely receptive to the need to protect INTELSAT from competition, and has limited its approval of separate systems to services not interconnected with the public switched network (which excludes public telephony) and to a limited number of circuits interconnected with the public switched network on every system. Upon the first consultation, INTELSAT found the 100 circuits interconnected with the public switched network per system would not cause the system significant economic harm. This figure was later raised to 1,250 circuits per satellite and, more recently, to 8,000 circuits per satellite.

Several U.S. companies currently operate separate satellite systems. PanAmerican Satellite Company Corporation (PanAmSat) based in Greenwich, Connecticut, operates three satellites with coverage of the American continent, Europe, and parts of Asia and Africa. Orion Satellite Corporation launched its first satellite (Orion 1) in 1994. Columbia Communications Corporation uses transponders on NASA's TDRSS satellite to offer civilian communication services. For more on separate satellite systems, see Chapter 5.

4.6 TOWARD THE TWENTY-FIRST CENTURY: MORE LIKE A GAZELLE?

"Rather than a dinosaur, INTELSAT is becoming more like a gazelle, quick and fleet of foot as it adapts to keep pace with the changing global communications environment." With these words, Director General Irving Goldstein announced the creation of an INTELSAT working group to examine the future of the organization in the new competitive international marketplace. The advent and multiplication of separate satellite systems and the continuous growth of cables have forced INTELSAT to come to terms with the end of its monopoly power and to face the need for radical change if the organization is to survive into the twenty-first century. The changes to be considered by the working group include converting INTELSAT to a private corporation, introducing multiple signatories, encouraging more direct access, permitting greater investment flexibility, and venturing into areas not traditionally exploited by the organization, such as specialized services and cables.

With these challenges in mind, the Assembly of Parties met in Venezuela in October of 1994 and approved a proposal to streamline procedures to amend the INTELSAT Agreement. Because under the current regime amendment of the treaty is extremely complicated and difficult to achieve, streamlining amendment procedures is the first step in changing the structure of the organization. The aim of this measure is to permit rapid implementation of the restructuring recommendations that the working party will recommend. Lack of agility has been one of the main criticisms leveraged against INTELSAT in the competitive environment. Although some parties do not believe that a complete overhaul of the organization is necessary to meet the challenges of the competitive environment, most agree that INTELSAT's management must become more commercial in nature. Most parties, with the exception of the United States, emphasized the need to preserve the intergovernmental nature of the organization and to reinforce the system's universal service and nondiscriminatory pricing obligations. These two points are of particular importance to smaller, remote, and less-developed nations, who fear their interests would be ignored in a commercial environment. The United States is in the minority in its belief that the best alternative is the complete privatization of the system. Its argument is that market competition and private investment are the best ways to improve the quality and variety of telecommunications services, reduce prices, and ensure flexibility and responsiveness to user demand. At the very least, at the behest of PanAmSat, the United States advocates a revision and elimination of all INTELSAT privileges and immunities, and the creation of a corporation affiliated with the organization that would manage the space segment commercially.

The United Kingdom is an enthusiastic proponent of multiple signatories. That country had to withdraw a proposal to permit multiple governors when the investment shares of two entities from the same country entitles each to a governor. The U.K. proposal faced wide opposition from developing countries and from Comsat. Those nations raised the concern that the Board of Governors could eventually be dominated by representatives from a handful of countries. However, this argument ignores basic commercial realities. Under the very terms of the INTELSAT Agreement, signatories are the representatives of the telecommunications operators, not of the governments of the countries under whose laws they are incorporated. That role is fulfilled by the parties. Governors, who are the representatives of signatories, represent the commercial interests of operators.

Under this regime, there is no guarantee, or even any serious likelihood, that two governors from the same country would represent identical or even compatible interests. Because telecommunications operators are increasingly international in their scope of operation, their interests can no longer be predicted based on national boundaries. Moreover, barring a second or third signatory from becoming a governor based on its nationality reinforces the second-class status of those signatories with respect to the original signatory. It is possible that for commercial and geostrategic reasons, second signatories may find it more advantageous to ally themselves with groups other than their country's. The current proposal leaves to the party the decision to designate one voting signatory and potential governor. Needless to say, this discretion works to the advantage of the incumbent, who will have a long-established cooperative relationship with the party, and preserves the status of the newcomer as an outsider. Moreover, an interesting problem emerges. Naturally, the voting signatory and the governor must represent and vote the interests of both signatories. This means that signatories must share information and cooperate in ways that may run afoul of their commercial interests or national antitrust laws. While INTELSAT itself is immune to antitrust prosecution by virtue of its intergovernmental nature, the privately controlled telecommunications operators are not.

France introduced a proposal to liberalize trade in investment interests. Currently, a signatory may purchase an investment share commensurate with its usage of the system. The French proposal would divorce usage and investment obligations and permit investors to trade investment shares freely. A signatory would be free to sell its share of the investment share that exceeds the minimum 0.05% necessary for continued INTELSAT membership. A cap of 150% of actual use was proposed to contain large investors such as Comsat and British Telecom. The Assembly of Parties also agreed to continue the relaxation of the significant economic harm assessment in authorizing separate systems, with a view toward eliminating it entirely at a future time.

At the same time, the U.S. government and PanAmSat, the largest U.S. separate system in operation today, are pressuring the organization to change and become more competitive by privatizing and perhaps by breaking up the organization into several regional operational companies following the model of the AT&T breakup. PanAmSat proposes a breakup of INTELSAT into three or four regional operating companies along the lines of its current service areas. During a House telecommunications subcommittee meeting, Rene Anselmo, PanAmSat founder and chairman, said that the best way to resolve the restructuring issues is by "ripping up" the INTELSAT Agreement.

INTELSAT has been the precursor of global connectivity and international communications. Its importance in developing modern telecommunications cannot be overstated. INTELSAT has also worked remarkably well as an international commercial experiment. It is a highly regarded and profitable enterprise. As global communications needs have evolved, cable systems and other satellite systems have begun providing INTELSAT with competition. The time has come to revisit INTELSAT's present and future role as provider of telecommunications services on a global scale. INTELSAT itself has undertaken a review of its role and other related issues. The outcome of this redefinition is uncertain. While competition has brought about the introduction of new services, it is not clear how universal service and nondiscriminatory access would fare under a fully competitive system. Smaller, economically deprived nations may come out as the losers in the process.

Notes

[1] Operators authorized by nonmember countries to use the space segment are known as *duly authorized telecommunications entities.*

[2] But, as will be discussed later, any application for a new Earth station that will use the system must be approved by INTELSAT.

[3] N. M. Matte, "International Arrangements for International Space Activities," in *Space Law, Development and Scope*, N. Jasentuliyana, ed., Praeger, 1992, p. 101.

[4] Agreement, Preamble.

[5] J. N. Pelton, *Global Communications Satellite Policy: INTELSAT, Politics and Functionalism*, Lomond Books, 1974, p. 42.

[6] Pelton [5], p. 41.

[7] Although Clarke was the first scientist to describe this phenomenon, it is a matter of simple Newtonian mechanics well understood since the seventeenth century.

[8] *The New York Times*, December 31, 1960, cited in Pelton [5], p. 48.

[9] UN General Assembly, 16th Session, Resolution 1721, Section P, cited in Pelton [5], p. 49.

[10] This proposal was introduced by Senator Kerf of Oklahoma on January 11, 1962. Pelton [5], p. 50.

[11] Communications Satellite Act of 1962, Pub. L. 87-624, 87th Cong., 2d Sess, 76 Stat. 419, enacted August 31, 1962, 47 USC 701 et seq.

[12] This proposal was introduced by Senator Estes Kefauver on February 26, 1962.

[13] U.S. Congress, House of Representatives, Committee on Government Operations, "Satellite Communications (Military-Civil Roles and Relationships)," Washington, D.C., 1965, pp. 90–91.

[14] Pelton [5], p. 54.

[15] Agreement Establishing Interim Arrangements for a Global Commercial Communications Satellite System, signed at Washington, August 20, 1964.

[16] Agreement Relating to the International Telecommunications Satellite Organization "INTELSAT," done at Washington, August 20, 1971, entered into force February 12, 1973. This agreement is known as the INTELSAT Agreement or the Intergovernmental Agreement. The terms will be used interchangeably throughout this chapter.
[17] Ibid, Operating Agreement.
[18] Several theories circulate. It has been suggested that the Soviet Union did not want to participate because the global satellite effort was led by the United States, because it was tantamount to United States imperialism, because it concerned itself with commercial uses of space while the Soviet Union was more interested in noncommercial educational uses, and because the Soviet Union, which was lagging behind in satellite technology, did not want its technological gap exposed in front of the United States and other nations. It is noteworthy that while the first U.S. communications satellite was launched in 1958, the USSR did not launch its first communications satellite, the Molniya I, until 1965. Pelton [5], pp. 55, 56.
[19] INTELSAT annual report for 1988–1989, p.45.
[20] Formally, the Agreement Relating to the International Telecommunications Satellite Organization "INTELSAT," done at Washington, August 20, 1971, entered into force February 12, 1973.
[21] Formally, the Operating Agreement Relating to the International Telecommunications Satellite Organization "INTELSAT," done at Washington, August 20, 1971, entered into force February 12, 1973.
[22] Pelton [5], p. 65. The INTELSAT I series was terminated after one launch (Early Bird has no siblings).
[23] INTELSAT has 130,000 international message toll service (IMTS) (or international telephone service) channels.
[24] This article provides for the construction and operation of separate satellite systems as long as they cause no economic or technical harm to INTELSAT (see Section 4.5).
[25] See Annex C to the agreement.
[26] Article XVI(b)(i) of the agreement.
[27] Public telecommunications services are defined in Article I of the agreement as fixed or mobile satellite communications services available for use of the general public.
[28] Article VIII of the INTELSAT Agreement.
[29] INTELSAT annual report 1994.
[30] Thus, the part of a signatory's investment share corresponding to specialized telecommunications services as defined in Article I of the INTELSAT Agreement is excluded from the computation for voting purposes.
[31] Malta joined in January 1995.
[32] In 1994, this amounted to approximately US$900,000.
[33] In 1990, requests for greater shares did not offset requests for lower shares. INTELSAT extended the period to request greater shares until the requests balanced out.
[34] M. S. Snow, The International Telecommunications Satellite Organization (INTELSAT), Nomos Verlagsgesellschaft, Baden-Baden, 1987, p. 94.
[35] Snow [34], p. 68.

5 Separate Satellite Systems

As we saw in Chapter 4, international communications via satellite are generally carried by the INTELSAT system. INTELSAT is an international consortium that owns and operates satellites for civilian communications. At the time INTELSAT was created, it was granted a monopoly over satellite communications because of the high cost of satellite construction and launch. INTELSAT was also meant as a means to advance world peace by promoting telecommunications. It has unique universal service obligations and provides its services to users at averaged prices independent of the actual costs of operating a route. The pricing scheme of INTELSAT is a form of subsidy from richer nations to poorer nations, and its viability depends on heavy use of the most lucrative, least cost routes.

INTELSAT is a cooperative whose members are the telecommunications operators of the member countries (the signatories). The statutorily created U.S. signatory is Comsat Corporation. At its inception, INTELSAT was the only satellite system capable of providing and authorized to provide international communications, but several of the countries that negotiated the original INTELSAT Agreements wanted to leave open the possibility of constructing national or regional satellite systems to serve their own needs. The INTELSAT Agreement therefore contains an escape clause to the grant of exclusivity in international communications. That escape clause is found in Article XIV, which allows for the construction of separate satellite systems for international communications if they do not cause a threat of harmful interference or economic harm to the system. Satellites used for domestic communications must not cause technical interference to the system.

Any company or country desiring to construct a satellite system, whether for international or domestic communications, must first ensure that it will not cause technical or economic harm to the INTELSAT system. The process of obtaining a certification that no harm will result from construction of a new satellite system is known as *consultation*. In addition, new satellite systems must undergo coordination at the ITU to ensure that they will not cause technical interference to existing or projected systems.

As satellite technology advanced and communications needs increased, constructing separate regional or national satellite systems became less a dream and more a pressing reality. Before constructing new systems, however, countries had to find a way to reconcile those systems with their obligations under the INTELSAT Agreements.

In the United States, the policy of INTELSAT monopoly has two exceptions: the separate systems policy and the transborder policy, both of which will be discussed in the following sections. Each evolved as a result of applications filed with the FCC seeking to establish alternative arrangements for international communications, and the FCC is currently considering revising and simplifying them. Direct access to the INTELSAT system by entities other than Comsat is also addressed.

5.1 SEPARATE SYSTEMS AND INTELSAT

The Executive Branch of the United States government saw the advent of separate satellite systems as a step forward for competition. It believed that users, especially large, sophisticated business customers, would benefit from the introduction of new services not offered by INTELSAT, such as large intracorporate networks particularly tailored to the customer's needs. At the same time, competition from separate systems would force INTELSAT to introduce new and better services at a lower price.

As we saw in Chapter 4, INTELSAT is a cooperative whose pricing structure indirectly subsidizes the telecommunications needs of poorer members by charging every user its share of the average cost of operating the system as a whole rather than the cost of the particular route used. This results in charges higher than costs for heavily used (or thick routes), such as New York–London, and charges dramatically lower than costs on rarely used (or thin) route, such as Bombay–Pointe Noire (Congo). In this context, it is hardly surprising that the major opponents of separate systems were developing nations, who feared that the introduction of competition on thicker routes would divert revenues from the system as a whole and result in higher overall prices or in a revision of the rate averaging mechanism to reflect costs on every route. If the thin routes—on which developing nations depend—were priced based on cost, the services would become prohibitively expensive for those nations.

To allay those fears, in the U.S., the Executive Branch restricted use of separate systems to communications not interconnected with the public switched network. Since the bulk of INTELSAT revenues are derived from public telephony (IMTS), they would be minimally affected by separate systems. In addition, in a booming telecommunications market, it seemed clear that the survival of INTELSAT as designed was not compromised by such separate systems, since any business lost to them would be compensated for by an increase in the global volume of demand.

Developing nations also worried that separate systems would claim the best orbital slots, thus compromising the goal of "equitable access" to the GSO under the ITU agreements. Rather than compromising equitable access, the United States claimed, the experience developed by separate systems operators would help developing nations serve their future communications needs. For example, the introduction of competition in domestic satellites made possible technological advances that reduced the necessary separation between domestic satellites from 5 to 2 degrees [1]. Developing nations could then take advantage of the technological advances resulting from separate systems when the time came to modernize and develop their own systems.

In the United States, separate systems were seen by some as a potential threat to international trade and to the privileged position of Comsat within INTELSAT. Some commenters believed that separate systems would have an adverse impact on international trade and would reduce U.S. participation in INTELSAT and consequently its influence over the organization. In addition, they believed separate systems would reduce the importance of INTELSAT as a buyer of U.S. aerospace products, would encourage the development of other regional and global systems that would buy equipment from other countries rather than the United States, and would lead to more reliance on submarine cables. The executive counterargued that the superiority of U.S. products, not any lobbying efforts on its part, was responsible for the fact that U.S. firms were awarded the almost the totality of contracts by INTELSAT. Today, however, in response to international pressure, INTELSAT purchases equipment and contracts launch services from firms of other nations as well (such as the French Ariane or the Chinese Long March). U.S. firms are still successful in competing against foreign firms in getting contracts for satellite technology and launch contracts abroad. Ford Aerospace was a contractor for both ARABSAT and the French satellite system, and Hughes Aircraft was the supplier for Indonesia's PALAPA. On the other hand, competition would bring down prices for consumers. International lines were overpriced, and their cost was not related to distance or any other objective criterion. A line from New York to London was priced at more than twice as much as a line from Los Angeles to New York, even though the distance and cost are comparable.

Finally, some parties raised national security concerns. The Department of Defense seeks to ensure the survivability of international communications by ensuring there is a certain redundancy of routes and mix of service options. Separate satellite systems would enhance national security by duplicating routes and multiplying the number of Earth stations capable of international communications.

At the same time, there is no doubt that the major client base of separate systems is the North Atlantic routes. Separate satellite systems do not have universal service obligations. They only seek to serve the most lucrative routes for the largest corporate customers. The only way this can be accomplished is by diverting traffic that would otherwise be carried by the INTELSAT system. The disappearance of this increased traffic cannot but result in higher average costs for thin-route users and for telephone subscribers in general. The INTELSAT system has met most of the domestic and international communications needs of developing countries, thus making scarce orbital slots available for the domestic and international needs of the United States.

By taking telecommunications discussions out of the international forum, separate systems compromise the friendly mood of consensus developed by INTELSAT. This consensual atmosphere has been a rare political feat. INTELSAT truly provides affordable state-of-the-art communications service to all nations in a nonpolitical context and it has been an unqualified success. Separate systems take the planning and discussion away from this unpoliticized cooperative forum. On the other hand, some players find the INTELSAT decision-making process slow and cumbersome.

5.2 THE LONG ROAD TO SEPARATE SYSTEMS

In the United States, the licensing of separate systems was not an easy process. The first applications were filed with the FCC in 1983 [2]. Shortly thereafter, the Department of State and the Department of Commerce sent a joint letter to the FCC asking it to refrain from taking any final action on the applications until the Executive Branch had examined the impact of separate systems on the national interest and foreign policy of the United States. In late 1984, President Reagan signed Presidential Determination 85-2 declaring that separate satellite systems were required in the national interest within the meaning of Sections 102(d) and 201(a) of the Communications Satellite Act [3]. While authorizing the creation of separate systems, the executive letter contained two restrictions on their operation: they must provide service only through the sale or long-term lease of transponders for communications not interconnected with the public switched network, and they had to undergo INTELSAT Article XIV(d) consultation and be authorized by at least one foreign administration before they could initiate service.

Two letters were also exchanged between Secretary of Commerce Malcolm Baldridge and Secretary of State George Schultz [4] stating that the Executive Branch must clarify its position on the flexibility of the pricing structure of cost-based access to INTELSAT (also known as *direct access*; for more on this topic, see Section 5.5). Meanwhile, a special Senior Interagency Group on International Communication and Information Policy led by the Departments of Commerce and State [5] presented a paper to the FCC reviewing the issue of separate satellite systems and explaining the position of the Executive Branch. This paper was entitled "A White Paper on New International Satellite Systems," but is generally known as the "White Paper." The paper reviewed U.S. policies to determine whether and under what circumstances separate satellite systems would be consistent with U.S. law and international obligations and in the public interest. The White Paper concluded that separate systems were compatible with the international obligations of the United States and with domestic laws and international treaties as well as in the public interest, but only if limited to the provision of "customized" services. By this term, the White Paper excluded public switched services from the scope of services separate systems may offer. The White Paper noted the growth in the telecommunications market and pointed out that competition already existed in the form of submarine cables. Other countries were already planning regional systems whose footprints would encompass at least part of the American continent. The Executive Branch believed that the restrictions placed on separate system with respect to public telephony would suffice to prevent major economic harm [6] to the INTELSAT system because it would avoid diversion of its core revenues, derived from international public telephony.

On January 4, 1985, the FCC initiated a rulemaking about the construction and operation of separate satellite systems [7]. A number of INTELSAT parties and signatories voiced their concern that separate systems would result in economic harm to INTELSAT. In view of the presidential determination, the FCC concluded that separate systems should be authorized because they would result in substantial benefits to users in the form of more service options at lower prices without resulting in major economic harm to INTELSAT. The FCC thought that the new services would increase the size of the telecommunications market, as frequent users such as financial institutions and data processing services would have access to more services and create more demand. In the rulemaking, users commented that INTELSAT did not offer an adequate range of services at reasonable prices. Moreover, the restrictions placed by the U.S. government on separate system operators were unique. No other country in the world had gone to such lengths to avoid harm to the INTELSAT system. To protect the system, the FCC imposed certain restrictions in addition to those mandated by the Executive Branch. It prohibited separate systems from acting as common carriers and extended the no-interconnection and long-term sale and lease restrictions to resellers and users of the facilities, but defined "long-term" as at least one year. It also clarified that no interconnection could be made at any point in the network by any carrier or user at the central office or through a private branch exchange (PBX).

Initially, therefore, separate systems could only be used for sophisticated business services not interconnected with the public switched network. The restrictions soon began to erode. In 1986, the FCC authorized provision of intercorporate communications and occasional-use television over separate systems [8]. In 1989, the Board of Governors of INTELSAT determined that it would not receive economic harm from interconnection of up to 100 circuits per separate satellite system to the public switched network. In 1990, PanAmSat filed a petition to further modify the separate system limitations. The Executive Branch began an exhaustive review and concluded that separate satellite systems should be permitted to offer international private-line circuits interconnected to the public switched network and that any restrictions on circuits interconnected with the network should be eliminated by 1997. The FCC soon issued an order announcing the changes, permitting interconnection, and eliminating the long-term contract requirement [9]. To manage the flood of applications that followed the availability of 100 circuits per system for public telephony, the FCC allocated the circuits among the applicants. By 1992, INTELSAT had determined that up to 1,250 64-Kbps equivalent circuits per satellite for the provision of public telephony would not cause the system economic harm. In December 1994, the Board of Governors announced that use of up to 8,000 circuits per satellite for public telephony would result in economic harm to INTELSAT. A letter from the U.S. Executive Branch is likely to be issued soon, but it is doubtful that it will have a significant impact, since the restriction on public telephony will be lifted in January 1997.

5.3 WHAT YOU CAN DO OVER SEPARATE SATELLITE SYSTEMS: CURRENT U.S. LAW

Separate satellite systems may not operate as common carriers themselves, but they can offer capacity to common carriers for the use of their customers. Under current law, end users can obtain any service from separate systems. Common carriers may offer enhanced services, private lines, and a limited amount of public telephony. Private lines may be interconnected with the public switched network provided that persons other than the customer and its employees and subcontractors do not use the private-line circuit to access the general public [10]. There is no restriction on the number of private lines that can be connected to the public switched network, but the lines may not be resold to provide public switched services. Until recently, services provided over separate satellite systems were subject to previous INTELSAT consultation on a country-by-country basis. Today, they are subject to consultation on a system-by-system basis.

5.4 THE TRANSBORDER POLICY

The FCC licensed the first commercial U.S. domestic satellites in 1973 to provide service exclusively within U.S. territory. However, because of the configuration of the satellites, their area of coverage (or *footprint*) exceeded the limits of the United States and spilled over into adjacent countries. Because the systems were licensed as purely domestic, the licensees could not offer any kind of international service. In 1977, however, several video service providers' networks applied for authority to receive satellite feeds directly from ANIK, the Canadian national satellite system. Because of the prohibition on international service, the applicants had been forced to receive the feeds through terrestrial links from Canada into a location in the United States, and then distribute them to their affiliates using the domestic U.S. satellite system. This arrangement made little sense, since the member broadcasting stations were technically capable of receiving the programming directly. Moreover, the indirect transmission also resulted in increased costs and duplication of facilities. The FCC found itself in the uncomfortable position of having to determine whether granting the applications would be in the public interest. On the one hand, because of the configuration of the satellites and the cost of indirect routing, the arrangements proposed by the applicants was clearly sensible. On the other, it had not received any guidance from the Executive Branch as to the scope of the obligations of the United States under the INTELSAT Agreements.

In 1980, a number of domestic nonvideo satellite providers also applied for authority to provide international service incidental to their domestic coverage. Satellite Business Systems (SBS) filed one of the first applications to use a domestic satellite to provide nonvideo international communications in 1980. The company had customers with branch offices and affiliated companies in Canada who wanted to have access to integrated private network services. SBS sought permission to utilize its domestic bird, whose footprint covered most of Canada, to offer those services between the United States and Canada [11]. SBS explained that the service was both practical and economical, and that any other alternatives, such as using terrestrial circuits or satellite double-hops, would be significantly more cumbersome and costly. Because at the time the application was filed neither the FCC nor the Executive Branch had developed a policy concerning international use of satellites outside the INTELSAT system, SBS proposed a complicated lease-back arrangement whereby it would lease capacity on its satellite to INTELSAT, which would then lease the capacity back to SBS through Comsat.

On July 23, 1981, the Department of State issued a letter to the FCC setting forth the executive's position on transborder transmissions. This letter is known as the "Buckley letter" after the undersecretary of state for security assistance, science, and technology. After examining the INTELSAT Agreement, the Department of State concluded that authorizing certain transborder transmissions over domestic satellites was not contrary to the obligations of the United States under the INTELSAT Agreements, but would be in the national interest of the United States under "[c]ertain exceptional circumstances." These circumstances include "where the global system could not provide the service required" or "where the service planned would be clearly uneconomical or impractical using the INTELSAT system" [12].

Of course, any use of a domestic satellite for international service still requires Article XIV consultation with INTELSAT. However, the Buckley letter also stated that international service could be offered after obtaining clearance from INTELSAT by a two-thirds majority in the Board of Governors *or* the support of the governments of both the United States and the country to which service would be inaugurated.

After receiving the Buckley letter, the FCC examined the applications to determine whether the services they proposed would be in the public interest. Comsat opposed the grant of the applications, claiming that they would cause immeasurable economic harm to the INTELSAT system, but this argument was summarily dismissed for lack of evidence.

Under the criteria of the Buckley letter, in every application for transborder services using a domestic satellite, the FCC must determine (1) whether existing or planned INTELSAT facilities are capable of providing the services in question, and (2) whether use of those facilities would be uneconomical or impractical. The FCC disposed of the video and the data applications on the same day.

In the video cases, the FCC found that although INTELSAT-proposed facilities would be capable of providing the service, the cost would be much higher, since the programming, which was already being carried on the ANIK birds, would have to be duplicated and rerouted to the INTELSAT system. This approach would be costly and inefficient. Moreover, to the extent the service was already available, cost considerations had precluded the use of the INTELSAT system for U.S.-Canada communications in favor of cheaper terrestrial circuits. In the case of Canada, the three-hop mechanism (U.S. Comsat-INTELSAT-ANIK) was nonsensical given that direct access was possible technically. The FCC found that the public interest would be served, since the new services would strengthen the ties of friendship between the United States and Canada and permit more U.S. viewers to receive Canadian broadcasts, especially educational and sports programming. With regard to the data applications, the FCC found that the new arrangements would also be more efficient and less costly for similar reasons.

The scope of the transborder policy was never clearly defined. The Buckley letter merely mentioned use of "domestic satellites" to serve "nearby countries." The FCC's decisions authorized only incidental transborder services over domestic satellites. Between 1981 and 1987, the FCC approved well over 100 transborder applications [13]. In 1986, Teleport International Ltd. and American Satellite Company applied for authority to provide two-way service between the United States and Jamaica. Because all previous nonvideo applications had been for either Canada or Mexico, the Teleport request raised a new issue: whether the requirement that transborder transmissions be "incidental" to a domestic service meant that they could only be an extension of a purely domestic service offering, or whether it could encompass cases in which the transmission was incidental to the footprint of the domestic satellite. The FCC had never before interpreted the scope of its restriction that the international service provided be incidental, whether to a domestic service or a domestic satellite.

The FCC granted the application after finding that although the service could be offered through the INTELSAT system, it would be uneconomical to do so. Comsat sought review of this decision before the court of appeals [14]. Quoting from earlier decisions, the court sided with the FCC and declared that transborder service could be incidental to a domestic satellite rather than to a domestic service. "We have made clear that the primary use of domestic satellites is for domestic service and that any use of these satellites for service to Canada, Mexico, the Caribbean or Central and South America should be viewed as incidental or secondary" [15]. However, the court also found that mere cost savings would not satisfy the "uneconomical or impractical" prong of the Buckley letter. Some sort of demonstration that the service provided would require duplicative facilities is needed.

It is important to remember that unlike the separate systems policy, the transborder policy does not now and never has prohibited the offering of voice services through the public switched network. This has been the main feature distinguishing the two policies. In a recent rulemaking, the FCC has proposed to abolish the distinction between the transborder and separate systems policies and regulate all FSS providers under the same scheme by eliminating the requirement that transborder transmission be "incidental" to a domestic satellite.

5.5 THE ISSUE OF DIRECT ACCESS

As we saw above (see also Chapter 4), any carrier or user desiring to use the INTELSAT system must access it through the designated signatory for that country. In the United States, Comsat Corporation, a statutorily created private corporation is the signatory. Since INTELSAT was created, Comsat has served as the U.S. signatory, the only U.S. investor in the system, and the sole supplier of access to the system. The international carriers, in turn, were the only providers of end-to-end service to customers. In many other countries, the signatory is the national PTT (e.g., France Télécom is the signatory for France, and Telia A.B. for Sweden). In others, it is the branch of the government that regulates telecommunications (e.g., in Argentina, the signatory is the Comisión Nacional de Telecomunicaciones (CNT), the Argentinean equivalent of the FCC). The signatory collects the anticipated space segment needs of its customers and makes all the necessary arrangements with INTELSAT to obtain the capacity. It then makes the space segment available to carriers for a price that includes a markup, or a profit margin. In some countries, as a matter of national law, only the signatory may own Earth stations.

This was the case in the United States until the 1980s. Comsat's customers were constrained to use Comsat facilities for their international transmissions.

Initially, Comsat was only a *carrier's carrier*, meaning that it could not provide service directly to end users. An end user who wanted to access satellite capacity for its communication needs had to obtain it indirectly from Comsat through a common carrier who could provide end-to-end service (such as AT&T). However, in 1982, the FCC allowed Comsat to offer service directly to noncarriers [16] and to engage in other, non-INTELSAT and non-INMARSAT lines of business in competition with the carriers. Carriers began worrying that the ability to offer end-to-end service to their customers would put Comsat in a competitively advantageous position compared to them.

In 1982, the FCC initiated a proceeding to consider the possibility of introducing direct access for U.S. carriers. Once Comsat was allowed into the carriers' traditional area of business, the carriers feared that having to purchase access from Comsat at a markup would mean that they would not be able to price their services competitively. They asked the FCC to allow them cost-based (referred to as direct access, regardless of whether the carrier is using its own Earth station or one owned by Comsat) to INTELSAT to remove this potential handicap.

Carriers defined direct access as "a form of access—by whatever name—which would allow...[them]...to obtain circuits at or near the same price that Comsat pays for the same facilities" [17]. They also noted that "[s]uch cost-based access would allow the carriers to compete on roughly the same basis for the provision of INTELSAT space segment to end users" [18]. Moreover, they argued, the prospect of earning a return on their investment would create incentives to maximize their use of the INTELSAT system and to utilize it more efficiently. In support of their position, carriers pointed out that not only signatories, but also the Soviet Union and users in other countries that were not members of INTELSAT could obtain access directly from the organization at the INTELSAT utilization cost (IUC) [19] rate (then, US$390 per month per half-circuit), while they were forced to pay a markup.

Thus, under U.S. law, the term *direct access* merely means that a user may make direct investments in the organization and obtain the capacity directly from INTELSAT without the intermediary of Comsat and therefore at the same price paid by Comsat. Two distinct forms of direct access were studied. The first was a capital lease under which Comsat would continue to make a capital investment in the organization, but would cease bundling the Earth station tariff and the space segment tariff. Carriers would be able to purchase Earth station facilities freely, from anyone they chose. They would lease the space segment from Comsat at cost and pay it a periodic ministerial fee to cover its administrative costs incurred in maintenance of the segment and participation in INTELSAT activities as signatory. The second was an indefeasible right of use (IRU) option under which the carriers would not obtain management or control rights in the space segment but would enjoy other benefits normally associated with ownership. They would be able to invest directly in the organization and collect the return distributed by INTELSAT and make direct payments to the organization. In return, they would pay Comsat for its share of INTELSAT's operating costs and a pro rata share of its administrative costs for maintenance of its corporate headquarters and its participation in international consultation meetings as the signatory for the United States.

Comsat opposed direct access on both statutory and economic grounds. As a statutory matter, according to its interpretation of the Communications Satellite Act, Comsat was, without a doubt, the only entity authorized to participate in the ownership of INTELSAT. Comsat also opposed direct access on economic grounds. As an economic matter, direct access would not permit it to recover its fair return for its investment, to which it is statutorily entitled, in order to attract new capital and compensate its investors. Moreover, direct access would prevent Comsat from recovering all the costs associated with being the U.S. signatory, such as the cost of participating in operational, financial, and technical planning meetings, and would severely curtail its research and development capabilities.

The FCC rejected the arguments of the proponents of direct access, finding that it would not produce significant efficiencies or cost savings for the end users. The FCC also found that direct access would not advance either intramodal (satellite-to-satellite) or intermodal (satellite-to-cable) competition. Moreover, it feared that permitting direct access would shift managerial control of Comsat from the organization to AT&T, its largest customer (which then utilized 90% of the space segment leased to carriers by Comsat). If allowed direct access, AT&T would be able to control Comsat's investment decisions and divert utilization from cables toward satellite in order to collect the return paid by INTELSAT.

Separate satellite systems permit direct access. Any carrier may purchase capacity directly from them without the intermediary of Comsat. In 1992, the Board of Governors of INTELSAT voted to permit direct access to the system by authorized users. As of 1995, two countries, Chile and the United Kingdom, permit direct access to the system. There is some question as to whether INTELSAT will permit direct access users to opt to become second signatories for their countries (see Chapter 4).

5.6 RECENT DEVELOPMENTS

In the Spring of 1995, the FCC undertook a review of its separate systems and transborder policies with a view to eliminating the distinction between them. As the economy becomes increasingly global in scope, businesses increasingly demand "one-stop shopping" and integrated services for their communication needs. The restrictions in both policies are only regulatory encumbrances. The

FCC has proposed to eliminate the transborder policy in its entirety and subsume it in the separate systems policy. Under the proposed new rules, every domestic satellite would be permitted to provide international service within its footprint, and every separate system licensed for international service would be allowed to offer domestic services within the United States.

There is no doubt that separate satellite systems will continue to be a crucial force. When the restriction on public telephony is lifted in 1997, they will provide additional competition to INTELSAT for INTELSAT's core business. The effects of this competition on INTELSAT cannot be ignored. The former monopoly is acutely aware that unless it reacts to the new competitive challenges, its days may well be numbered. At the same time, separate satellite systems are providing innovative services at competitive prices without a signatory intermediary. Those services can often be accessed with cheaper and smaller Earth stations than Earth stations approved by INTELSAT.

Notes

[1] L. A. Caplan, "The Case for and Against Private International Communications Satellite Systems," *Jurimetrics J.*, Winter 1986, p. 190.

[2] Orion Satellite Corporation, File No. CSS-83-002-P, 11 March 1983; and International Satellite, Inc., File No. CSS-83-004-P(LA), I-P-C-83-073, August 12, 1983.

[3] 47 U.S.C. 701(d) and 721(a)(6), which authorize the creation of separate satellite systems if "required in the national interest."

[4] See letter from Malcolm Baldridge, secretary of commerce, to George P. Schultz, secretary of state (November 30, 1984), and letter from George P. Schultz to Malcolm Baldridge (December 20, 1984), cited in 101 FCC 2d 1054.

[5] But composed of representatives of the Departments of State, Justice, Defense, and Commerce; the Offices of Management and Budget, Science and Technology Policy, Policy Development, and the United States Trade Representative; the National Security Council; the Central Intelligence Agency; the U.S. Information Agency; the Board of International Broadcasting; the Agency for International Development; and NASA.

[6] In making a determination of economic harm, INTELSAT considers factors such as the amount of traffic that will be diverted from the system and the types of services (domestic or international) concerned.

[7] *In the Matter of Establishment of Satellite Systems Providing International Communications*, 100 FCC 2d 290 (1985), Notice of Inquiry.

[8] 61 RR 2d 649 (1986).

[9] *Permissible Services of U.S. Licensed International Communications Satellite Systems Separate from International Telecommunications Satellite Organization (INTELSAT)*, 7 FCC Rcd. 2313 (1992).

[10] *Alpha Lyracom d/b/a PanAmerican Satellite*, 9 FCC Rcd. 1282 (1992).

[11] *Satellite Business Services*, 88 FCC 2d 195 (1981).

[12] Letter from undersecretary of state for security assistance, science, and technology to Chairman Fowler, dated July 23, 1981, reprinted at 88 FCC 2d 259, 287 (commonly known as the "Buckley letter").

[13] *Comsat v. FCC.*

[14] *Comsat v. FCC*, 64 Rad. Reg. 2d 414 (D.C. Cir. 1988).

[15] *Western Union Tel. Co.*, File No. 1144-DSS-P/LA-84, Mimeo No. 6929 (FCC September 11, 1985), quoted in *Comsat v. FCC.*

[16] *Authorized User Policy*, 90 FCC 2d 1394 (1982), commonly known as "Authorized User II."

[17] Networks Comments at 6. The Networks favored the adoption of a direct access policy, and they submitted joint comments. They are the American Broadcasting Company (ABC), the National Broadcasting Company (NBC), and the Columbia Broadcasting Company (CBS).

[18] Ibid.

[19] The IUC is based on INTELSAT's estimate of its annual revenue requirement per unit of space segment.

Becoming a Player: The International Section 214 Process

6

We now turn to domestic regulation of telecommunications common carriers and how those carriers are licensed and regulated by the FCC.

The Communications Act enjoins all carriers to seek permission from the FCC before extending lines to provide telecommunications service or, once service has commenced, to discontinue it. Section 214 of the Act says, "No carrier shall undertake the construction of a new line or of an extension of any line, or shall acquire or operate any line, or shall engage in transmission over or by means of such additional or extended line, unless and until there shall first have been obtained from the Commission a certificate that the present or future public convenience and necessity require or will require the construction."

For domestic communications, the requirement to obtain previous FCC authority has been eliminated as part of the FCC's efforts to reduce the regulatory burden on carriers. Today, a carrier's Section 214 authority to provide domestic service is subsumed in its facilities license (satellite, microwave, or cable), while no specific authority or license is needed to be a domestic reseller.

The FCC has retained the previous licensing requirement in the international arena, however. Before offering service to its customers or to the public, a carrier must obtain authority from the FCC. Because it is Section 214 of the Communications Act of 1934 that requires carriers to seek FCC permission, that license is known as "Section 214 authority," and the application that precedes it is commonly known as a "Section 214 application."

Not all 214 applications were created equal. The FCC has subdivided the international arena into several categories. A carrier may be a common carrier or a private carrier. Common carriers may be facilities-based or resellers, dominant or nondominant. They may also be affiliated with foreign carriers. Each of those factors has an influence on the application process, its grant, and on the final terms under which a carrier may serve its customers.

This chapter will explain the difference between the several categories of carriers and the attempts of the FCC to develop a coherent and predictable way of regulating international carriers.

6.1 COMMON CARRIERS VS. PRIVATE CARRIERS

The term *common carrier* is not defined in the Communications Act of 1934. Title II of the Act, which regulates common carriers, tautologically states that a common carrier is "any person engaged in common carriage for hire, in interstate or foreign communication by wire or radio." The definition of common carrier was established by common law, and it was based on old common law notions. Black's Law Dictionary defines common carrier as "[a]ny carrier required by law to convey passengers or freight without refusal if the approved fare or charge is paid *in contrast to private or contract carrier.* One who holds himself out to the public as engaged in business of transportation of persons and property from place to place for compensation, and who offers services to the public generally" [1]. Traditionally, common carriers have traded obligations such as universal service for limited liability in case of loss or destruction of the goods they carried. This tradeoff applies in telecommunications as well. Common carriers were carriers offering their services to the public at large. In exchange for ceding control of the goods to the carrier, customers and because of the control carriers had over the means of transportation and their safety, those carriers became *quasipublic* entities subject to a higher standard of care akin to that of an insurer. In return, however, their monetary liability was limited to reimbursing the price of the service provided. The quasipublic nature of common carriers has been used to justify a higher level of regulation as well. The same holds true for telecommunications entities. The crucial difference is that while private carriers are virtually unregulated, common carriers must reckon with a copious body of regulation.

For purposes of FCC regulation, the current definition of common carrier was established judicially in 1975. (See *National Association of Regulatory Utility Commissioners v. FCC,* 525 F.2d 630 (D.C. Cir. 1976), the "NARUC case.") After reviewing the history of common carrier law at some length, the court opined that the distinction between common and private carriers was one of marketing. Thus, a common carrier is one who "undertakes to carry for all people indifferently." However, the court pointed out, this does not mean that the product or service must be one of wide appeal. "One may be a common carrier though the nature of the service rendered is sufficiently specialized as to be of possible use to only a fraction of the total population." This does not mean customers may not be turned away under any circumstances, but that the carrier will not make individualized decisions based on subjective factors. Rather, the decision will be based on whether it has enough capacity available, or whether the carriage is consistent with the carrier's usual business. That approach need not be written down as company policy. It merely has to be the carrier's general business practice.

Private carriers are free to pick and choose their customers and to enter into individual fee agreements with them. On the other hand, they may not interconnect most of their services to the public switched network. Either a common carrier regime or a private carrier may be used for video transmissions. Under either regime, a carrier may, at its option, transmit through the INTELSAT system or a separate system, such as PanAmSat, Columbia, Orion, or INTERSPUTNIK. The choice of a satellite system has legal as well as technical implications. INTELSAT transmissions tend to be of better quality. On the other hand, separate satellite systems offer more competitive pricing and obviate the need to use Comsat as an intermediary.

Legally speaking, common carrier regulation is far more burdensome than private carrier regulation. For instance, the FCC has created a distinction among common carriers between dominant and nondominant (for more on this distinction, see Section 6.4 below). That distinction does not exist in the world of private carriers. To operate internationally as a common carrier, an entity must first obtain a Section 214 authorization from the FCC. This authorization is country-, service-, and satellite system–specific, although it is possible to request every country, service, and satellite system in one application. Private carriers are not subject to this requirement, and their Earth station licenses alone are sufficient to authorize them to provide service.

Common carriers are required to prepare tariffs that must be filed with the FCC for public inspection for some time before they may go into effect. These proposed tariffs must be just and reasonable and may not be unduly discriminatory. Tariffs may be challenged by any member of the public as "unjust," "unreasonable," or "unduly discriminatory." Such complaints may take up to two years to resolve and may result in an award of damages against the carrier found liable. Common carriers are obligated to offer their services at the tariffed rate to anyone requesting it, even to their competitors. Private carriers provide their services at their discretion, since they are not obligated to serve any member of the public on demand. Their charges are negotiated in private contracts that need not be filed with the FCC as public documents, and they need not be the same for each customer or type of customers.

As we will see in Section 6.6, common carriers using their own Earth station facilities may not be owned or controlled by foreign nationals or by a company in which foreign nationals own or vote 20% or more of the stock directly or 25% indirectly [2]. Instead, foreign-controlled carriers are limited to transmissions by cable and to leasing the uplink and downlink services of U.S.-owned and controlled carriers. Private carriers, on the other hand, are not subject to these foreign ownership restrictions.

In summary, a private carrier may offer many of the same services as a common carrier, except for public telephony and services interconnected with the public switched network. It may operate by using the same satellite and cable systems. However, private carriers are not subject to the same degree of regulation, including foreign ownership restrictions, as common carriers. They need not apply for and obtain prior authorization to serve new points or to use new satellites or cables, need not hold themselves out to the public indiscriminately, and need not file tariffs.

6.2 BASIC SERVICE OR ENHANCED SERVICE

Enhanced services are also known as *value-added* services. They generally consist of services in which a computer interacts with information (e.g., voice, data) provided over telephone lines. The FCC's rules define enhanced services as "services...which employ computer processing applications that act on the format, content, code, protocol or similar aspects of the subscriber's transmitted information; provide the subscriber additional, different, or restructured in- formation; or involve subscriber interaction with stored information" [3]. Enhanced services are not regulated. No authorization, application, or notification is required to provide them, and they are not tariffed. All other communications services are basic and therefore regulated. Examples of enhanced services include fax store and forward, videotext, electronic mail, and voice mail. When it is not clear whether a service is basic or enhanced, the FCC will examine the nature and function of the service rather than its form to make a determination.

6.3 SECTION 214 SERVICES: CABLES, INTELSAT, AND SEPARATE SYSTEMS

A carrier may offer any service over transoceanic cables or over the INTELSAT system. Because of the original monopoly granted to INTELSAT, however, only certain services may be offered through separate satellite systems. For more information about this topic, see Chapters 4 (INTELSAT) and 5 (Separate Satellite Systems).

INTELSAT offers myriad services, but their definition does not necessarily coincide with the FCC's. Although the FCC recognizes international business service (IBS) as a category of service and permits carriers to apply to provide it to their customers, it does not, for example, recognize intermediate data rate (IDR) service. That INTELSAT service supports many communications applications, including videoconferencing and public telephony. However, the FCC requires that applicants seeking to provide this service specifically list and describe the individual services they will provide (e.g., switched voice, private lines, fax).

A separate satellite system is any international satellite system other than INTELSAT. As explained in earlier chapters, in order to protect the economic viability of INTELSAT, certain restrictions were placed on the use of separate systems. Thus, a carrier may offer video services, public switched telephony, private lines, data transmissions, ATM applications [4], private lines interconnected with the public switched network at either or both ends, or any other service through cables or through INTELSAT. The decision of which to use will be based on technical considerations. While cable generally affords better quality and is believed by some to better support certain applications [5], it is subject to frequent breaks that interrupt service and are costly to repair. Satellites offer greater versatility and worldwide coverage at no additional cost. In some instances, they may be the only available option. For example, it is impossible to reach Tanzania, Madagascar, or Malawi by cable, but INTELSAT offers satellite coverage of those countries.

Separate satellites systems may only be used to provide services not interconnected with the public switched network, with certain exceptions. Initially, public switched or interconnected services could be offered through separate systems. In 1990, however, INTELSAT determined that the use of separate satellite systems up to 100 64-Kbps equivalent circuits per system would not cause significant economic harm to INTELSAT. The U.S. government soon followed suit by authorizing carriers to acquire up to 100 circuits per system [6]. In November of 1992, INTELSAT changed its determination of "no significant economic harm" from 100 circuits per system to 1,250 54-Kbps equivalent circuits per satellite. The FCC followed by implementing further changes pursuant to instructions from the Executive Branch by finding that increasing the number of available circuits would serve the public interest. In 1994, INTELSAT modified its economic harm finding once more. Today, it considers that interconnected transmissions up to 8,000 circuits per satellite as not causing it economic harm. The U.S. government, however, has not yet authorized carriers to obtain that number of circuits on separate systems.

Separate satellite systems may be used for any kind of communication, whether interconnected or not, to countries that are not members of INTELSAT, except those against which the United States has imposed an embargo (such as, until very recently, North Korea and Cuba). However, as the world changes and more and more countries become members of INTELSAT, this exception is becoming increasingly meaningless. Currently, it only applies to a handful of African countries (such as Angola) and to some of the former Soviet Republics and Yugoslav Republics (such as Uzbekistan, Kirghizia, Montenegro, and Macedonia—but this list is not exhaustive). Carriers may offer any service, whether interconnected with the public switched network or not, to those countries. They may also offer value-added services, which are unregulated, to any country though separate systems. For countries that are members of INTELSAT, certain additional restrictions apply. In the United States, pursuant to FCC rules, each separate satellite system may only be used for certain countries [7].

6.4 DOMINANT OR NONDOMINANT

Dominant regulation is an increased form of regulation imposed on some carriers for competitive reasons. The first carrier to receive dominant regulation was, of course, AT&T.

The classification of a carrier as dominant or nondominant has important regulatory implications. The regulation of nondominant carriers has been streamlined; that is, they are subject to a less stringent form of regulation. For example, where tariffs are concerned, dominant carriers must file their rates 45 days before they are to go into effect, while nondominant carriers need only file them 14 days before they become effective.

Originally, all international carriers received a higher level of regulation. In 1985, the FCC conducted a rulemaking proceeding to decide whether to relax those requirements to certain carriers who, because of their lack of market power, could not affect the price of international telephone service. The FCC proposed to streamline the tariff and facility regulations of those carriers in order to reduce the regulatory burden on them so as to better accomplish its statutory mandate to bring to U.S. users "rapid, efficient, nation-wide and world-wide wire and radio communications services with adequate facilities at reasonable charges." In 1979, the FCC had revisited its domestic rules and found that competition in that market was strong enough to ensure that market forces provide protection to the public from unreasonably high rates and undue discrimination. The FCC thus introduced the concept of *forbearance*. Under that concept, the FCC ceased to apply certain sections of Title II of the Act to certain nondominant companies (basically everybody but AT&T) and exempted those carriers from filing tariffs. Forbearance was later struck down by the courts, and all carriers must now file tariffs.

With respect to international carriers, the FCC went through a lengthy proceeding to determine whether there was enough competition in the international arena to justify lessened regulation. First, it concluded that there were three product markets: international voice (IMTS), international television service, and non-IMTS, such as telex, telegram, private lines, and IBS, an INTELSAT service offering. The FCC based this distinction on the notion of supply and demand substitutability and cross-elasticity of demand between those services. Then it considered the proper market definition for telecommunications services and concluded that, because of the need to conclude operating agreements with foreign carriers in order to offer service, the appropriate geographic market is best defined on a country-by-country basis. The FCC then used those parameters to determine whether certain carriers lacked market power, and could therefore receive lessened regulation without affecting the level of competition in the market. The FCC found that no carrier in the non-IMTS market qualified as dominant, and that only AT&T should be classified as dominant in the IMTS market, mainly because of its overwhelming market power and the fact that it owned the majority of the existing undersea cables.

The FCC recognized that the task of defining the product market would necessarily yield artificial distinctions. Two main factors were considered: demand substitutability and supply substitutability. Demand substitutability is a subscriber's ability and willingness to switch between services. It is affected by factors such as quality, price, availability, geographic coverage, and marketing. Thus, two products are substitutes if a rise in the price of one will lead subscribers to switch to the other. The FCC found that IMTS, which constituted 70% of revenues and 80% of traffic, was not substitutable for non-IMTS services such as private lines. Although there is substantial voice traffic over private lines, most private lines are used for the transmission of data. Those transmissions do not suffer from time and language constraints, and therefore they are a distinct product from IMTS.

Supply substitutability is defined in terms of entry barriers. It exists when a service provider may shift its resources from one service to another to reflect its perceived demand for the service and the ability to make profits in excess of its opportunity costs. The most significant barriers to entry into the international market are the need to obtain a Section 214 authorization and the need to conclude an operating agreement with a foreign correspondent. The FCC also found that only a few carriers offer IMTS service, while the great majority of them provide some sort of non-IMTS service. It also found that non-IMTS operating agreements are easier to conclude than IMTS agreements.

A carrier's geographic market is defined by its ability to provide service in an area and by a customer's ability to receive it. The FCC found that every country is a separate geographic market because a separate operating agreement must be concluded with a foreign telecommunications entity to serve it. In addition, because of a scarcity of facilities, it is not easy for a carrier to shift its resources to serve a different country when demand increases.

Market power is defined as "the power to control prices or exclude competition" in a particular geographical area. A firm is dominant if it can drive its competitors out of the market by lowering prices without causing entry of new competitors into the market.

By applying this analysis, the FCC found that AT&T was the only U.S. dominant firm in the IMTS market, since it was the only provider of that service to a great number of countries. In routes where AT&T had competitors, the FCC found that their market penetration was insufficient to warrant nondominant regulation of AT&T. In addition, it determined that AT&T enjoyed a market advantage because of its worldwide coverage, such that subscribers were more likely to choose it over others. All other U.S. IMTS providers were found to be nondominant. On the non-IMTS front, however, the FCC found that no carrier had market power. Even AT&T only had less than 10% of the market and was unable to affect competition.

Initially, the FCC maintained full regulation on all routes for all foreign-owned carriers. This decision was seen as the most efficient way to persuade foreign operators to grant operating agreements to U.S. carriers on a nondiscriminatory basis. A carrier was defined as foreign-owned if 15% or more of it was controlled directly or indirectly by foreign interests, or if an employee, agent, or representative of a foreign telecommunications entity sat on its board of directors. The FCC changed this policy in 1992.

In 1992, the FCC redefined foreign ownership to mean control and ruled that only foreign-affiliated carriers would be regulated as dominant. A U.S. carrier is now said to have a foreign affiliation if it controls, is controlled by, or is under common control with a foreign carrier. Foreign affiliation is the functional equivalent of foreign ownership. Under the new policy, carriers are only regulated as dominant on the route where their correspondent is their affiliate. However, they may be regulated as nondominant if they can demonstrate that they lack the ability to discriminate against nonaffiliated U.S. carriers because their foreign affiliate lacks market power and the ability to control bottleneck facilities. Bottleneck control includes a legally protected monopoly or a monopoly in fact due to the absence of a regulatory regime that ensures competition. Today, carriers that have no affiliation with another carrier in the destination country are presumptively nondominant for that route. Carriers who are affiliated with a monopoly in the destination country are presumed to be dominant, and carriers whose affiliate is not a monopoly will receive close scrutiny by the FCC to determine whether their foreign affiliate has the ability to discriminate against nonaffiliated U.S. carriers. However, the FCC has yet to find a foreign-affiliated carrier nondominant on a route where its affiliate is a correspondent. Examples of foreign-affiliated carriers that have received dominant regulation include ATN [8], Telefónica of Spain, and AmericaTel. All U.S. international resellers are presumptively treated as nondominant regardless of foreign affiliations.

The tariffs of nondominant carriers are presumptively lawful and need only be filed 14 days before they are to go into effect. Dominant carriers must cost-support their tariffs, which must be filed 45 days before they are to go into effect. Dominant carriers must specifically apply for new circuits, while nondominant carriers may add any circuits they wish once they have received authority to serve a particular country. Dominant carriers must also file quarterly revenue and traffic reports within 90 days from the end of each calendar quarter.

6.5 RESELLER OR FACILITIES-BASED

The FCC has never truly defined the difference between resale and facilities-based carriers. However, that distinction is crucial, since it impacts the type of Section 214 application a carrier must file, the level of review by FCC staff, and how soon a carrier may expect a grant. Resellers, also known as *aggregators* when they do not handle their own billing, are generally understood to be those carriers that do not operate their own network or Earth stations and who buy capacity in bulk from other carriers, which they then resell to their customers. This option makes sense for carriers that have a level of traffic that does not justify building a network. Using the network of other carriers gives them the flexibility they need to route traffic and permits them to avoid having to conclude operating agreements with foreign companies. Instead, they buy a service package from a facilities-based carrier into which the costs of transmission and the accounting rate are already factored in. Resellers may have their own switch or may merely resell the switched service of another carrier (in which case they are called *switchless resellers*). Both do their own marketing and generally bill the end customer themselves. However, their traffic is considered the underlying carrier's for most FCC reporting purposes.

Domestic resale began when AT&T decided to win back the customers it had lost as a result of the divestiture and the introduction of competition. AT&T cut its rates dramatically and relaxed its bulk billing practices. In 1976, the FCC had found that permitting unlimited resale of domestic common carrier lines was in the public interest, and prohibited facilities-based carriers from limiting resale by their customers. A number of new companies emerged that offered significant discounts to end users, especially business customers. AT&T removed its tariff restriction on the resale of IMTS and private lines in 1982 and 1986, respectively. AT&T acted as a market leader, and most other carriers soon removed their resale restrictions as well. With respect to international service, however, unlike service in the domestic arena, the removal did not follow an FCC order, and some carriers have retained the restriction.

Soon a few companies began applying for Section 214 authority to resell international lines. In 1991, the FCC found unlimited resale of IMTS and private lines in the public interest. In making this determination, it reviewed the results of its domestic resale order and concluded that increased resale had been instrumental in increasing demand and reducing cost for telecommunications services and in eliminating the possibility of price discrimination. The FCC found that a similar opening of international resale would also increase demand and reduce prices to bring them more in accordance with the real costs of providing the service, and would reduce collection rates [9] while exerting downward pressure on accounting rates [10]. When considering the possibility that resale may harm the underlying carriers [11], the FCC pointed to the experience in domestic data indicating that the traffic had not been diverted from the facilities of the carrier, and that any price reduction had been more than offset by a volume increase [12]. However, since foreign correspondents usually bargain with U.S. carriers from an advantageous position, the FCC introduced certain safeguards. In order to protect U.S. carriers from undue pressure from foreign carriers, the FCC permitted them to resell private lines only to countries that afford similar resale opportunities in the reciprocal direction.

The presence of similar resale opportunities must be evidenced by either a statement that the FCC has already determined that equivalent resale opportunities exist, or a demonstration that they do in fact exist. Equivalence might be shown by demonstrating similar licensing, tariffing, and other terms and conditions of service. At a minimum, this must include equal treatment for U.S.-based carriers in terms of the factors referred to above.

In November 1992, the FCC made its first such finding when it certified that Canada offers equivalent resale opportunities. Two carriers, fONOROLA Corporation and EMI Communications Corporation, had filed applications to offer private-line resale between the United States and Canada. EMI and fONOROLA demonstrated that several decisions of the Canadian Radio-Television Commission (CRTC) had progressively liberalized the resale market, and that there was an open entry policy for resale, regardless of nationality. fONOROLA pointed out that of the approximately 80 resellers then operating in Canada, at least 10 were U.S.-based [13]. Moreover, any conditions placed on operation in the Canadian market, such as the obligation to maximize use of Canadian facilities, applied equally to all operators without discrimination. All resellers had the ability to interconnect with the public switched network under the same terms and conditions. The FCC therefore authorized those two carriers to resell private lines and established that Canada offered equivalent resale opportunities. However, in order to prevent this finding from permitting other carriers to circumvent the international resale policy and undermine its goal of reducing collection rates and exerting downward pressure on accounting rates, the FCC prohibited the two companies from routing U.S.-overseas traffic through Canada. All subsequent Section 214 authorizations were also granted with the expressed condition that the companies could only carry U.S.-Canada traffic over the resold lines. In addition, because fONOROLA was a foreign-owned company, it was subject to dominant regulation.

In 1994, the FCC made an equivalency finding with respect to the United Kingdom. The FCC found that the United Kingdom does not restrict the entry of U.S.-based telecommunications carriers into its market [14], that all carriers received the same treatment regardless of nationality, and that the United Kingdom permitted the free interconnection of private lines to the public switched network. Moreover, the United Kingdom has a policy of only permitting resale of private lines to countries it has found to offer similar opportunities to U.S. companies [15]. Thus, the FCC concluded that the terms and conditions to provide international resale in the United Kingdom are equivalent to those in the

United States. However, the FCC granted the applications subject to the condition that the United Kingdom find that the United States offers equivalent resale opportunities to British companies. On October 20, 1994, the president of the Board of Trade designated the United States as a country offering equivalent resale opportunities, and the companies were authorized to commence service. Similar applications have been filed concerning Finland, France, Germany, Australia, and Sweden.

The resale applications of IMTS carriers and providers of private lines not interconnected with the public switched network now benefit from a simplified processing and review regime called *streamlining* [16]. The applications of resellers to countries for which an equivalency finding has been made benefit from this process. Streamlining has been codified in Section 63.12 of the FCC's rules [17]. Those applications are placed on Public Notice for 30 days, during which any interested party may oppose them. If no one opposes them, and if the FCC finds no objection to the application after cursory review, they are automatically granted without further FCC action on the 45th day after acceptance, and the reseller may begin operating on the 46th day. The Public Notice of the grant will appear on the next day FCC actions are published, and that publication will constitute the Section 214 authorization of the carrier.

To benefit from streamlining, the applicant may not be affiliated with the common carriers whose capacity it seeks to resell, or initially with a carrier in the country of destination. Affiliation is defined as mutual or common control. Initially, resellers of private lines could not benefit from this process unless they provided service to a country in which the FCC had determined that equivalent resale opportunities existed for U.S. carriers. Both the foreign affiliation limitation and the equivalent resale opportunities limitation were abolished for resellers of simple private lines (not interconnected with the public switched network) in 1992. However, the FCC retained the equivalent resale opportunities requirement for private lines interconnected with the public switched network.

Applications for facilities-based service may only be granted pursuant to a written order specifying their terms and conditions. A carrier is generally understood to be a facilities-based carrier when it uses its own network, space segment, or Earth stations to provide service. Thus, a carrier that purchases capacity from INTELSAT or from a cable consortium would be a facilities-based carrier. Recently, IDB Communications asked the FCC to clarify that definition by stating that a facilities-based carrier is one that acquires the maximum interest in a cable or satellite allowed by the law. The FCC tentatively declined to revise its definition and reaffirmed that a facilities-based carrier is one who has an ownership or IRU interest in a U.S. half-circuit on a submarine cable or a satellite or if it leases a U.S. half-circuit from Comsat or from a cable provider [18].

While the functional distinction between a facilities-based carrier and a reseller may not be clear to the end user, there are, as we have seen, profound differences between them in regulatory terms. Facilities-based carriers enjoy more freedom and flexibility in offering their services to the public. In return, they must contend with a longer and more cumbersome application process.

6.6 THE SPECIAL PROBLEM OF FOREIGN OWNERSHIP

Two types of authorizations are required to operate a satellite communications common carrier. The first is a Section 214 authorization, which gives a company authority to carry traffic to certain countries. This type of authorization is granted under Title II of the Communications Act, which regulates common carriers. Title II imposes no restrictions on foreign ownership of telecommunications companies. In addition, a company that plans to use its own teleport must obtain a license from the FCC for its antennas. Because Earth stations use the radioelectric spectrum, they are regulated under Title III of the Communications Act. Title III restricts the amount of foreign ownership a common carrier licensee of an Earth station may have to 20% if held directly and 25% if held indirectly out of national security concerns.

The legislative history of the foreign restrictions of Title III of the Communications Act indicates that Congress was concerned about the threat to national security if foreign interests controlled the broadcast media in times of war. This perception came about in the thirties, with memories of World War I still fresh on American minds.

Section 310(b) of the Communications Act directs the FCC not to grant a "broadcast or common carrier" radio license to "any corporation of which any officer or director is an alien or of which more than one-fifth of the capital stock is owned of record or voted by aliens or their representatives" [19]. The Act also prohibits indirect ownership of more than 25% of a licensee by foreign interests "if the Commission finds that the public interest will be served by the refusal ...of such license" [20]. In practical terms, this means that entities exceeding those limits may not become the licensees of common carrier Earth stations.

The same restrictions apply to partnership interests and even to limited partners. The FCC has found that limited partnership interests are ownership interests within the scope of Section 310(b) because the principles announced there apply to all business forms [21]. Limited partners hold equity in the company and may be involved in the management. Since the purpose of the foreign ownership restriction is to protect the means of communication from undue foreign influence, aliens may not use the limited partnership structure to obtain interests they would not otherwise be permitted to have. Since foreign nationals may not be officers or directors of U.S. licensees, limited partners must be "properly insulated" from management of the common carrier. Although the FCC places the burden of demonstrating that a foreign limited partner is properly insulated, it does not prescribe a method of proving proper insulation. Carriers are free to present whatever evidence they deem probative to rebut this presumption. The same approach applies to indirect ownership interests. As we saw in Section 6.1, foreign ownership limits do not apply to private carriers or to the acquisition of interests in cables. Foreign-controlled carriers may obtain Section 214 authorizations, but they may not hold Earth station licenses. The solution is obvious. Foreign-controlled carriers lease the uplink and downlink services of U.S. Earth station licensees and then connect the traffic to their own networks.

The FCC has been much more flexible with indirect ownership interests. The Communications Act merely enjoins the FCC to deny or revoke a license where foreign interests hold more than 25% of the stock of the parent of a licensee, or if foreign citizens constitute more than 20% of the directors of the parent company *if the public interest so requires it.* Since the Communications Act gives the FCC considerable discretion in this area, the FCC has found that several companies whose parent corporations exceed the limits comply with the Act. The FCC has permitted foreign ownership interests of 60% and more in common carrier Earth station licensees. While doing so, it has stressed the fact that common carrier licensees have no control over the contents of their transmissions, and that the national security concern is lessened. In all the cases where interests exceeding the limits have been exceeded, the applicants were from English-speaking countries whose close cultural and economic ties with the United States made them a natural ally of this country.

In early 1995, Senator Larry Pressler, chairman of the Senate Committee on Commerce, Science, and Transportation (also responsible for telecommunications matters) introduced a draft telecommunications reform bill that would lift the foreign ownership restrictions on a reciprocal basis. Under Senator Pressler's model, after the United States Trade Representative (USTR) has determined that a particular country offers an open market and open entry to U.S. telecommunications companies, it will lift the restrictions as to nationals of that country. If the USTR's finding is later reversed, the country would be decertified, and the licenses granted under this regime would be revoked. This model would apply to both directly and indirectly controlled companies.

The FCC is also making attempts at reform. In March of 1995, it issued a *Notice of Proposed Rulemaking* announcing a revision of the international Section 214 process with respect to foreign interests, and proposing to adopt a reciprocity standard. It also proposed the same standard for indirect foreign ownership of Earth stations. Because of the clear mandate of Section 310(b)(3) of the Act, the FCC has no discretion to modify its rules with respect to direct ownership interests.

Foreign ownership comes into play in Section 214 proceedings with the issue of foreign affiliation. Although there are currently no restrictions on who may obtain a Section 214 authorization, the terms of that authorization and the level of review the application will receive are significantly affected when a

U.S. carrier is affiliated with a foreign carrier. A U.S. carrier is an affiliate of a foreign carrier if it "controls, is controlled by, or is under common control with a foreign carrier" [22]. When a carrier is affiliated with a foreign carrier, its applications must be reviewed by the full FCC. It will be regulated as dominant on the route on which its affiliate is also a carrier unless it can demonstrate that competition exists in the affiliate's country. The level of competition must be such that the affiliate cannot discriminate against unaffiliated carriers, because the number of participants in the market and their respective market shares is such that the affiliate has no market power and that the affiliate has no bottleneck control of essential facilities necessary for the origination or termination of calls. It bears noting that no foreign-affiliated U.S. carrier has yet been found to be nondominant under this standard. Carriers that have no affiliation in the destination country and switched-service resellers will be presumptively considered nondominant for that route.

6.7 FILING AN APPLICATION: CHARTING THE SECTION 214 PROCESS

Now that we know the basic rules of the game, it is time to apply them. Under current FCC rules, an applicant must submit an original and five copies of its Section 214 application to a post office box in Pittsburgh, together with a filing fee of $705 in the form of a check made payable to the FCC. In Pittsburgh, Mellon Bank handles all fee collections for the FCC. If the fee is correct, Mellon Bank will forward the materials to the FCC in Washington, D.C. This process may take up to a week. Once the application is received in Washington, it is sent to the International Bureau for review. If a cursory review of the application reveals no major problems, it is then placed on Public Notice as accepted for filing. Applications that contain major errors (or are "fatally flawed," in FCC parlance) are returned to the applicant, who is given the option to resubmit them after the pertinent corrections have been made.

A Public Notice is a document published by the FCC regularly listing applications accepted for filing or actions taken on applications. The date of a Public Notice starts the clock on a period of time in which any interested party may oppose an application or petition for reconsideration of a grant. Appearance on a Public Notice of the acceptance of an application for filing triggers a 30-day period in which any party may file an opposition to the grant of the application or to any aspect thereof (for example, a party may ask that an applicant be regulated as dominant instead of as nondominant while not opposing the grant of the application generally). The applicant then has 10 days to file a response. For example, if the Public Notice of acceptance for filing of the application of International Telco, Inc., appeared on June 5, 1995, interested parties would have until close of business, July 6, 1995, to file their oppositions, and

International Telco, Inc., would have until July 17, 1995, to file its response. All in all, it takes approximately two weeks from the time an application is sent to Mellon Bank in Pittsburgh to the time it appears on a Public Notice.

As we saw earlier, unopposed resale applications that benefit from streamlined processing are automatically granted without FCC action 45 days after their acceptance for filing is announced in the Public Notices. Where those authorizations are concerned, although the authority to operate goes into effect on the 45th day (even if it is a Sunday or a holiday), the grant only appears in the next Public Notice (see Figure 6.1).

If in our earlier example, International Telco, Inc., filed an application to resell the service of another international carrier, and the acceptance for filing of the application appeared on a Public Notice on June 5, and if no one filed an opposition to the application, its authority to operate would be effective July 20, and if no Section 214 notices appeared on that day, the grant would be announced on the next day Public Notices appeared.

A Public Notice of a grant triggers a 30-day period in which any party that participated in the application process may petition the FCC for review of the grant, and an additional 10 days during which the FCC may decide to set aside the grant on its own motion. At the end of that 40-day period, a grant is said to be "final," since it is no longer subject to appeal. However, when a petition for review is filed, a grant remains in effect until a final decision has been reached. In our earlier example, if the next Public Notice came out on July 24, the grant would become final 40 days later, on August 31, 1995.

Nonstreamlined applications are granted when the FCC issues a document known as an Order and Authorization, specifying the conditions of the grant. The issuance of that document appears on a Public Notice (for an example, see figure). Under a new FCC policy, unopposed facilities-based Section 214 applications that do not present novel or complex issues will be processed within 30 days after the period to oppose them has expired. However, it is very difficult to predict with certainty how soon an application will be granted, since the FCC does not always stick to this schedule. For applications that present more complicated issues, such as a foreign affiliation, the process is somewhat more complex. After initial review, the attorney processing the application will refer it to his or her supervisor with a recommendation. When a foreign affiliation exists, the application must be considered by the full FCC, even if the affiliation does not concern the route applied for.

Under special circumstances, the FCC may grant an applicant special temporary authority (STA) to operate before its Section 214 application has been granted, or even before an application has been filed. These authorizations require a demonstration of extraordinary circumstances and are granted for a six-month period only [23]. They are rarely granted, except where humanitarian efforts are concerned, such as aid to refugees or emergency war relief. Grants of this type of authority are a matter of days, but if the service is to continue be-

FEDERAL COMMUNICATIONS COMMISSION USA

PUBLIC NOTICE

FEDERAL COMMUNICATIONS COMMISSION
1919 M STREET N.W.
WASHINGTON, D.C. 20554

53569

News media information 202/418-0500. Recorded listing of releases and texts 202/418-2222.

Report No. I-8045

April 26, 1995

OVERSEAS COMMON CARRIER SECTION 214 APPLICATIONS
ACTIONS TAKEN
(Formal Section 63.01)

The following applications for international Section 214 certification have been granted effective April 21, 1995, pursuant to the Commission's streamlined processing procedures set forth in Section 63.12 of the Commission's Rules, 47 C.F.R. Section 63.12. (All are resale of public switched service).

ITC-95-198	American Telephone & Fax, Inc.
ITC-95-199	International Long Distance Telephone Company, Inc.
ITC-95-200	Citizens Telecommunications Company, d/b/a Citizens Long Distance Company
ITC-95-201	Access Unlimited, Inc.
ITC-95-202	Pacific General Telecom, Inc. d/b/a Westar Network

The applicants listed above are authorized to provide international switched services by reselling the international switched services of other carriers as listed in their application, and only in accordance with all rules, regulations and policies of the Commission.

The following applications for international Section 214 certification have been granted effective April 24, 1995, pursuant to the Commission's streamlined processing procedures set forth in Section 63.12 of the Commission's Rules, 47 C.F.R. Section 63.12. (All are resale of public switched service).

ITC-95-209	Eastgate Communications, Inc.
ITC-95-210	CalTech International Telecom Corporation
ITC-95-211	Cellnet Telecommunications L.L.C.
ITC-95-212	Concourse Communications, Inc.
ITC-95-213	USN Communications, Inc.
ITC-95-215	National Telecom Management, Inc.

Figure 6.1 Public notice announcing the granting of several Section 214 applications.

The applicants listed above are authorized to provide international switched services by reselling the international switched services of other carriers as listed in their application, and only in accordance with all rules, regulations and policies of the Commission.

Provisions Pertaining to All Applicants

All of the applicants listed in this public notice shall file a tariff pursuant to Section 203 of the Communications Act of 1934, as amended, 47 U.S.C. Section 203, and Part 61 of the Commission's Rules, 47 C.F.R. Part 61, for the services requested in their application. The applicants shall file the annual reports of overseas telecommunications traffic required by Section 43.61 of the Commission's Rules, 47 C.F.R. Section 43.61. Further, the grant of these applications shall not be construed to include authorization for the transmission of money in connection with the services the applicants have been given authority to provide. The transmission of money is not considered to be common carrier service.

If an applicant is reselling services obtained pursuant to a contract, the applicant shall file publicly any contracts entered into with other carriers or a contract summary in accordance with Section 203 of the Communications Act, 47 U.S.C. § 203, and the Interexchange Order.[1] In addition, the services obtained by contract shall be made generally available to similarly situated customers at the same terms, conditions and rates.

To the extent that any of the above-listed U.S. carriers intend to provide international call-back services through the use of uncompleted call signaling, their authorization to resell international switched voice and/or data services to provide these services is without prejudice to, and is expressly subject to, any future action the Commission may take in VIA USA Ltd., et al., 9 FCC 2288 (1994), petition for reconsideration pending.

Petitions for reconsideration under Section 1.106 or applications for review under Section 1.115 of the Commission's Rules in regard to the grant of any of these applications may be filed within 30 days of the date of this public notice (see Section 1.4(b)(2)).

For additional information concerning this matter, please contact Janice Alston (202) 418-1493.

-FCC-

[1]See Competition in the Interstate Interexchange Marketplace, 6 FCC Rcd 5880, 5902 (1991) (Interexchange Order).

Figure 6.1 (continued)

yond the six months allowed under the rule, that application should be followed, if not preceded, by a full-fledged Section 214 application.

6.8 THE SPECIAL CASE OF CUBA

The Cuban Embargo Act of 1961, passed in the wake of Castro's takeover of Cuba, prohibited communications between the United States and Cuba. For a number of years, only AT&T was able to provide voice communications between the two countries under a special waiver. However, any monies for settlements with the Castro government had to be placed in a frozen account in the United States. In 1993, this account was worth an estimated US$80 million. In 1989, however, hurricane Hugo destroyed some of AT&T's microwave facilities used in its Cuba link. Thereafter, AT&T operated 14 circuits on an indirect route through Italy. Initially, AT&T was allowed to terminate 20,000 minutes per day in Cuba, but the Cuban government later limited that number to 20,000 minutes per month. An undersea cable between Miami and Cuba remained inactive because of the refusal of the Cuban government to permit its activation. In exchange, Cuba had asked for the release of the frozen funds, a position the U.S. government found unacceptable.

In 1994, it was estimated that 60 million calls were attempted, but only 1% of those calls were completed. This situation encouraged the emergence of pirate companies that used bypass arrangements to provide service to Cuba. Several of those companies advertised openly in the Spanish-speaking press and television channels of cities with large Cuban populations, such as Miami. The bypass arrangements used an 800-number to connect callers to the Canadian network. Since Canada had never signed the Cuban Embargo Act or a similar document, the call was then routed to Cuba through the facilities of a Canadian carrier. Some of those companies charged up to US$23 for a three-minute call.

In July of 1993, the U.S. government began cracking down on these operations, and the Treasury Department's Office of Foreign Asset Control (OFAC) issued letters to several carriers ordering them to cease operations immediately. At the same time, the U.S. government announced a relaxation of the embargo with respect to telecommunications under the Cuban Democracy Act of 1992. That document liberalized the provision of telecommunications services as a way "to seek a peaceful transition to democracy and a resumption of economic growth in Cuba" [24] by permitting those services as well as the construction of "[t]elecommunications facilities...in such quantity and of such quality as may be necessary to provide efficient and adequate telecommunications services between the United States and Cuba" [25].

On July 27, 1993, the FCC announced that it would begin accepting applications for service to Cuba pursuant to guidelines received from the Treasury

Department and the State Department [26]. For the first time, those guidelines allowed for direct payments to the Cuban government. However, in addition to the normal Section 214 application requirements, applications for service to Cuba had to comply with several Treasury Department conditions and be approved by the State Department, the Treasury Department, and the FCC. Those conditions are operability of the project within a year of authorization, construction limited to equipment and service necessary to provide the service, settlements of no more than US$1.20 and a 50/50 split, and no access to the AT&T frozen account. The Cuban government rejected the initial overtures of U.S. carriers and demanded an accounting rate of US$2.00 and release of the funds in the frozen account.

Several U.S. carriers soon filed applications, together with requests for STA. The initial applications were rejected because the Treasury Department found a surcharge of US$4.85 per call on collect calls imposed by the Castro government to be a violation of the Uniform Settlement Policy. However, the FCC permitted those carriers to amend their applications to reflect a new surcharge of no more than US$1.00 (the average surcharge rate). On October 5, 1994, the FCC granted five applications for service to Cuba for switched voice and private line after the carriers negotiated a new surcharge of US$1.00 with EMTELCuba. The Cuban government announced it would lift its restrictions soon thereafter, and service was initiated in late November of 1994. The approximately 200,000 access lines available in Cuba were soon swamped by callers from the United States who were able to dial their family members' numbers directly for the first time in over 30 years [27].

Applications for service to Cuba must be circuit-specific and must comply with the Treasury Department and State Department guidelines. However, all resellers that obtain capacity from facilities-based carriers may offer service to Cuba without seeking further authority if the underlying carrier holds a 214 authorization to serve that country.

In order to begin providing telecommunications services, a carrier must obtain a Section 214 authorization from the FCC. Carriers may be resellers or facilities-based, and they may be regulated as dominant or nondominant. In recent months, the FCC has been exploring ways to simplify the application process and to reduce the regulatory burden on carriers. Before it may provide a service, a facilities-based carrier will also have to obtain FCC authorization to acquire cable capacity or space segment for its operations. Resellers must conclude a contract with an underlying carrier and obtain Section 214 authority as well. The FCC is currently redefining its policy on foreign ownership policy. It is not yet clear whether the end result of this revision will be a regime more or less stringent than the current rules, which significantly limit foreign participation in Earth station ownership, but do not restrict ownership of communications companies that do not operate Earth stations.

Notes

[1] *Black's Law Dictionary*, abridged 5th edition, West, 1983 (emphasis added).

[2] Under Sections 310(b)(4) and (5) of the Communications Act, which applies only to radio transmissions.

[3] This definition is contained in 47 C.F.R. §64.702.

[4] See Chapter 1.

[5] Such as ATM, although a new INTELSAT service called Super-IBS will also support this application.

[6] See letter dated December 14, 1990, from Thomas J. Murrin, deputy secretary of commerce, and Lawrence S. Eagleburger, deputy secretary of state, to Alfred C. Sikes, chairman of the FCC.

[7] For example, PanAmSat may only be used for transmissions not interconnected with the public switched network between the United States and countries that are not members of INTELSAT, plus Anguilla, Argentina, Aruba, Australia, Austria, Azerbaijan, Bahamas, Barbados, Belgium, Bermuda, Bolivia, Brazil, the British Virgin Islands, the Cayman Islands, Chile, Colombia, Costa Rica, the Czech Republic, Denmark, the Dominican Republic, Ecuador, France, Germany, Greece, Guatemala, Haiti, Honduras, Hong Kong, Ireland, Italy, Jamaica, Luxembourg, Mexico, Monaco, Montserrat, the Netherlands, the Netherlands Antilles, New Zealand, Panama, Paraguay, Peru, Portugal, Romania, the Russian Federation, Spain, Sweden, Switzerland, Trinidad and Tobago, the Turks and Caicos Islands, the United Kingdom, Uruguay, Venezuela, and the former Yugoslav republics. It may be used for transmissions interconnected with the public switched network between the United States and Australia, Azerbaijan, Bahamas, Costa Rica, the Czech Republic, the Dominican Republic, Hong Kong, the Netherland Antilles, New Zealand, Panama, Peru, Portugal, Romania, the Russian Federation, and the United Kingdom. Other systems, such as Orion, Columbia, and the INTERSPUTNIK system are even more limited. However, as INTELSAT relaxes its economic harm policy, these restrictions are bound to disappear.

[8] See *Atlantic Tele-Network, Inc.*, 6 FCC Rcd. 6529 (1991).

[9] The price of a call to the telephone company's customer.

[10] The charges a telephone company must pay its foreign correspondent to terminate a call.

[11] Carriers that own the transmission facilities that the reseller is reselling.

[12] 70 RR 2d 160.

[13] *In re Applications of fONOROLA Corporation and EMI Communications Corporation*, 7 FCC Rcd. 7312 (1992).

[14] With the notable exception of international facilities-based telephony, reserved for British Telecom and Mercury only.

[15] As of the date of the order, those countries included Canada, Sweden, and Australia.

[16] 7 FCC Rcd. 7331, 7336 (1992).

[17] 47 C.F.R. §63.12.

[18] Foreign ownership NPRM.

[19] Section 310(b)(3).

[20] Section 310(b)(4).

[21] See *Wilner & Scheiner*, 103 FCC 2d 511 (1985).

[22] *In the Matter of Regulation of International Common Carrier Services*, 7 FCC Rcd. 7331, 7332 (1992).

[23] Section 63.04 of the FCC's rules governs STA under Section 214. For authorizations for emergency service, it requires showing that there is an "immediate need occasioned by conditions unforeseen by, and beyond the control of, the carrier." 47 C.F.R. §63.04 (1993).

[24] Cuban Democracy Act of 1992 Section 1705 (e)(1), Pub. L. 102-484-Oct. 23, 1992, 106 STAT 2575, 2576.

[25] Ibid., p. 2578.
[26] Public Notice of July 27, 1993, Report No. I-6831.
[27] There had been widespread concern that the ancient Cuban network would not be able to handle the predicted number of incoming calls. It was estimated that half of the 17 central office switches in Cuba were equipped with step-by-step switches and the other half with crossbar. The switches were either prerevolutionary or Eastern Bloc equipment.

7 The Rules of the Game: Tariffs, the ISP, and International Accounting Rates

In addition to obtaining operating authority from the FCC through the Section 214 process (see Chapter 6), international carriers must follow certain rules that are designed to protect consumers, preserve competition between U.S. carriers in international services, and ensure fair dealings with foreign administrations. They must file tariffs or contracts describing their services, file certain annual reports, and follow fairly strict rules in negotiating operating agreements with their foreign counterparts. In addition, they are now under an obligation to help the U.S. government lower intercarrier international rates (known as *accounting rates*). Those rates are seen by a number of foreign governments and carriers as an easy source of revenue and a form of subsidy from richer nations to poorer nations. Finally, there are some limitations on the services they may provide to certain countries, such as resale of international private lines and call-back arrangements, that are designed to help those countries preserve the revenue stream created by their telecommunications monopolies.

7.1 TARIFFS: WHO MUST FILE THEM AND HOW

As we have seen in previous chapters, the jurisdiction of the FCC extends only to interstate communications. The FCC may not regulate the terms of service, tariffs, or any other aspect of purely intrastate calling. For example, the FCC will have no jurisdiction over a call from Poughkeepsie, New York, to Manhattan. It will, however, have authority over a service provided between Poughkeepsie and Newark, or between Newark and London. The following rules apply only to common carriers as defined by the FCC (see Chapter 6). Because they are presumed to deal with more sophisticated business customers rather than the public at large, private carriers are not subject to the same constraints.

The first FCC rule affecting international carriers, and the one that most directly affects callers, is the obligation to file tariffs. A tariff is a list of the telecommunications services a carrier offers detailing the terms and conditions of the offering, such as the price per minute, the method of payment, and the grounds for termination of service. Under Title II of the Communications Act of 1934, tariffs may not be unjust, unreasonable, or unduly discriminatory [1]. This means that two similarly situated customers may not be treated differently, for example, by offering them different prices for the same service or different interconnection or service terms.

Every international provider of basic services must file a tariff with the FCC, which will place it in a public reference room where any member of the public may see it and copy it. It must also make the tariff available to any member of the public upon reasonable demand. Since enhanced services are unregulated, providers need not file tariffs for those services [2]. A copy of each tariff must be kept at the offices of the carrier and be available for inspection by an FCC official or any interested member of the public. In addition, an operator must be available to discuss the rates of the service free of charge to any person making a call. The operator must reveal the charges per minute upon demand and without charge to the user.

All international carriers, whether dominant or nondominant, must file tariffs. Dominant carriers must file their tariffs on 45 days' notice or 45 days before they can go into effect. The tariffs of dominant carriers must be cost-justified. This means that they must provide an economic basis for the prices they charge for their services for every tariff they file on a route on which they are dominant. For example, if AT&T prepares a new tariff for IMTS on its U.S.-Italy route and it intends to begin implementing it on July 16, 1996, it must file a copy of that tariff with the FCC no later than June 1, 1996. Other carriers have until that date to ask the FCC to suspend the tariff and investigate it as unlawful. The FCC may also suspend and investigate the tariff on its own motion.

Nondominant carriers, on the other hand, need only file their tariffs on 14 days' notice and need not provide cost justification for them. Thus, if LDDS wants to implement a new tariff between the United States and Italy on July 16, it must file it with the FCC by July 1, 1996. The tariffs of nondominant carriers are presumed legal, and they will not be investigated unless a competitor impugns their lawfulness. Any time a carrier changes the terms and conditions of its service (for example, every time it seeks to implement a price reduction, no matter how temporary), it must file a new tariff and give the appropriate notice. The new tariff will not become effective until the notice period has passed.

Tariff or Contract?

Not every carrier has to offer its services to every customer on a tariffed basis. Nondominant international carriers can offer service on a contract basis, generally to other carriers purchasing bulk capacity from them for resale to their customers. They are then permitted to negotiate special discounted rates and other arrangements. These special contracts, however, may not be exclusive. A copy of all special contracts or a summary containing a description of all important terms and conditions must be filed with the FCC, and the same general terms and conditions must be available to any other similarly situated customer who requests them. Failure to make the same terms and conditions available to other carriers may constitute undue discrimination and may subject the offending carrier to sanctions.

7.2 THE INTERNATIONAL SETTLEMENTS POLICY, OR PROTECTING YOUR COMPETITORS

U.S. carriers have an obligation to protect intercarrier competition. They may not enter into exclusive arrangements with each other or with foreign carriers. The FCC imposed this obligation because it feared that foreign monopoly carriers negotiating exclusive deals with U.S. carriers would abuse their position and threaten intercarrier competition in the United States, thereby destroying the main public benefit of competition: more options and lower prices for users. The International Settlements Policy (ISP), also known as the Uniform Settlements Policy (USP), was a response to this concern. It was created in the 1930s and has its roots in federal antitrust law.

Inextricably tied to the ISP is the problem of international accounting rates, which has puzzled and frustrated U.S. international telephone companies and the FCC for decades (see Section 7.7). In order to understand the ISP and the international accounting rates problem, a little historical background is useful. In the old days, there were only two telecommunications entities involved in any international communication: the PTT of the country where the call was originated and the PTT of the country where it was received. Each was compensated for its participation in completing the transmission. The telephone or telegraph company that originated the message billed its customer. The price paid by that customer is known as the *collection rate*. Naturally, the company then had to compensate the other PTT for terminating the call. The price paid by a PTT to another PTT for terminating a transmission originated by its customer is known as the *accounting rate*.

Unlike collection rates, which are published tariffs that every customer desiring service has to pay, accounting rates are agreed upon in private negotiations between carriers. When agreeing to open a new route, two carriers will sit at the negotiating table and discuss all the details of their relationship, including the technical facilities to be used, the modes and schedule of payment, routing, the price they will pay each other for terminating each other's transmissions (i.e., the accounting rate), and the division of that price (the settlement rate). For example, if France Télécom and Telefónica del Perú are negotiating to open a new route, they may agree that the accounting rate is $2.00 per minute, to be divided 50/50. This means that for every minute of traffic that France Télécom sends Telefónica del Perú, it will pay Telefónica del Perú $1.00. Likewise, Telefónica del Perú will pay France Télécom $1.00 for every minute France Télécom terminates for a Telefónica del Perú customer. At the end of the fiscal year, ideally, neither company must pay the other, since the traffic flow is symmetrical and the amounts even out.

This arrangement works to everybody's satisfaction as long as certain market conditions exist. In order to provide international service, the two carriers must necessarily cooperate. If no agreement is reached, the route cannot be opened and both parties lose. When only two carriers are present, each will have a strong incentive to reach a mutually satisfactory agreement whereby both win and neither loses. Another necessary market condition for the system to work is symmetrical traffic flows. Ideally, each carrier would send the other a number of minutes equal to the number of minutes it receives from that carrier so that the balance at the end of the fiscal year is close to zero. Note that no one is considering the interests of the subscribers or bargaining for lower prices on their behalf.

7.3 THE UNIFORM SETTLEMENTS POLICY: ITS ORIGINS

The preceding scenario was established at a time when there was a limited number of players in the world of telecommunications, and two entities of equal bargaining power sat at the negotiating table. As a model, this scenario is still viable as long as the market conditions described above are not disturbed.

Complications arise when there is more than one carrier per country on any route. The second carrier is a newcomer who must compete against the established PTT for market share. As the new market entrant, it has no market share and it must obtain customers in any way it can in order to gain market share. Since there is usually no substantial quality difference between a telephone call placed through two companies [3], the market newcomer must generally compete on price to gain customers. This presents it with a dilemma. On the one hand, it must offer low collection rates in order to attract callers. On the other hand, foreign PTTs have little incentive to negotiate with the newcomer unless it can offer them something too.

What the newcomer can offer is a more favorable deal to the foreign PTT, such as a higher accounting rate, or a division of tolls other than 50/50 in favor of the foreign operator. The newcomer has nothing to lose, since it begins the game with no market share and it needs routes to attract customers. Foreign monopolies, who know this, can and do play carriers against each other to obtain concessions and business advantages from either or both competitors. A monopoly sitting at the negotiation table with a second carrier has no particular incentive to open a route with that carrier. For the newcomer carrier, however, the route is a matter of life or death. The monopolist therefore has an unfair negotiating advantage. The practice of playing one carrier against another in order to extract concessions from one or both is known as *whipsawing.*

This term was coined by the FCC in 1936 in a now-famous case, *In the Matter of Mackay Radio & Telegraph Company, Inc. ("Mackay")* [4]. Mackay concerned the International System, a web of affiliated radio, landline, and cable companies that provided competitive international telegraph service. In the period between 1934 and 1936, it received 50% of the total revenue earned by U.S. companies providing international telegraph service. In 1936, Mackay Radio, a member of the International System, decided to install a new radio circuit directly connecting the United States with Oslo, Norway, to compete against R.C.A. Communications, Inc. In order to get the Norwegian administration to enter into an operating agreement with Mackay Radio, the International System offered to send all its unrouted traffic between the United States and Norway over the new radio circuit rather than over cables, and offered a higher accounting rate to Norway for the radio circuit than for the cable traffic. This arrangement would cause a loss of revenue to the International System as a whole and would virtually eliminate the U.S.-Norway cable traffic of another International System affiliate, but Mackay Radio was willing to sacrifice the other companies in order to gain market share.

Naturally, in addition to an agreement with the Norwegian PTT, Mackay needed Section 214 authority from the FCC to open a new circuit.

When the Section 214 application came before it, the FCC found that the new circuit would not generate additional traffic, but would merely shift the routing of existing traffic from cable to radio circuits. Since Mackay had agreed to a lower share of the tolls than that offered to the cable companies, the Norwegian administration would have a strong incentive to send its traffic through the Mackay circuit, since that would increase its revenue [5]. Mackay in turn would benefit to the extent the increased traffic volume made up for the reduction in its collections from the new settlement rate.

The FCC added that to expect Norwegian PTT officials to send traffic through the Mackay circuit under those circumstances was merely to expect them to be good businessmen and to safeguard the national interest.

Moreover, the arrangement set a dangerous precedent for other U.S. carriers, since "the maximum concession which any American...company makes to a foreign...administration becomes the minimum which any competitor can make to that administration [and it] is impossible to foresee the ultimate maximum of concessions which a company will make in a desperate effort to get or retain traffic" [6]. The FCC denied Mackay a license to construct the new circuit, and the court of appeals upheld the decision, noting that "Congress has recognized that competition between carriers may result in harm to the public as well as in benefit" [7].

In reaction to Mackay, the FCC adopted its Uniform Settlements Policy in the 1930s. The USP requires all U.S. international carriers offering similar services to the same foreign point to do so under identical accounting rates, settlement rates, and divisions of tolls. It also requires carriers to accept only their proportion of return traffic from the foreign correspondent. Carriers were thus prohibited from negotiating exclusive rates and conditions of service with foreign carriers. The principle of *proportionate return* directs each carrier to accept traffic from its correspondent only in the same proportion as it receives it from the correspondent. For example, if France Télécom receives 30% of all its traffic from the United States through Sprint, it may not, and the operating agreements with other carriers must say it will not, send more than 30% of its U.S.-bound traffic through Sprint.

The uniformity requirement is meant to ensure that both service providers receive fair compensation for their services and to prevent the foreign administration from whipsawing its U.S. correspondent, for example, by promising it an increased amount of traffic in return for a distribution of settlements more favorable to the foreign carrier. In the scenario described above, the foreign entity has both the ability and the incentive to play one U.S. carrier against the others in order to extract concessions for its own benefit, for example, by promising Sprint 50% of its outbound traffic in exchange for a lower accounting rate, a division other than 50/50, or better technical interconnection arrangements.

No rule is airtight, and the USP is no exception. In 1980, the FCC issued a policy statement in connection with a request to reduce the telex accounting rate between the United States and the United Kingdom, reaffirming the need for the USP, but stating that the public interest may warrant waivers of the policy in certain cases. The public interest factors that may warrant such waivers were promoting lower collection rates, improved service, and increased competition. However, since negotiations and discussions between carriers typically occur before waiver requests are filed, whipsawing can escape the supervision of the FCC and masquerade under an arrangement that seemingly advances the goals promoted by the USP.

The FCC did not initially establish a formal waiver procedure or a timetable for action on waiver requests. On the other hand, it initiated a series of meetings with industry representatives to discuss the problem of whipsawing and search for possible solutions. The meetings did not prove entirely fruitful, and the FCC revisited its USP. In addition, the new level of competition in voice services (which until not long before had been reserved to AT&T) introduced a new dimension to the problem. The time was ripe for a new definition of the obligations of international common carriers toward each other. The FCC revised its policy in light of the new market developments and introduced the ISP as it functions today.

7.4 MODERN REVISION OF THE INTERNATIONAL SETTLEMENTS POLICY

In 1986, the FCC revisited the USP to consider reforming certain aspects [8]. Noting that the goal of the policy was to protect consumers and U.S. carriers and not uniformity per se, the FCC renamed the USP the ISP. The FCC reiterated the harmfulness of whipsawing, which by reducing the revenues of U.S. carriers prevents them from lowering collection rates, thus harming the interests of consumers by threatening competition within the United States.

In response to a Notice of Proposed Rulemaking, commenters underscored the need for continuation of the USP in an increasingly competitive world. They pointed out that foreign monopolies were becoming increasingly sophisticated and able to take commercial advantage of U.S. carriers by whipsawing them in novel ways. At the same time, they advocated a revision of the USP and criticized the FCC's application of its waiver process as taking into account only one factor: whether the proposed variation would increase the total revenues of U.S. carriers. Some carriers advocated freeing small carriers (those without market power) from the requirements of the USP to permit them to be more competitive, but that suggestion was ultimately not adopted.

The FCC issued a reminder that the USP's purpose is to protect U.S. ratepayers, and not U.S. carriers, against abuses by foreign companies that would prevent them from enjoying the benefits of the introduction of competition in the U.S. market. In this context, a careless revision of the USP, not to mention its elimination, would shift the benefits of competition from the U.S. ratepayer to the foreign PTTs through whipsawing. The FCC therefore reaffirmed the policy while modifying certain procedures. Most notably, it significantly simplified the waiver process and streamlined it. The revisions addressed four areas: transit traffic, basic voice, enhanced services, and general procedural requirements for waivers.

7.4.1 Transit Traffic

The term *transit traffic* refers to traffic routed through the United States for commercial or technical reasons, but which neither originates nor terminates in the country. For instance, for technical or commercial reasons (better facilities, lower accounting rates), a call from Italy to Peru may be routed through the United States rather than through a direct satellite link. Neither the person making the call nor the person receiving it will be aware of this arrangement. The call merely flows through U.S. facilities. Transit traffic must be distinguished from indirect traffic, which is switched through an intermediate third country, but is either originated or terminated in the United States. For example, when communications between the United States and Cuba were not permitted, some unlicensed carriers connected calls originated in the United States to numbers in Cuba by routing the calls through Canada or other countries that had a direct communications link with the island.

The FCC did not apply the USP to transit or indirect traffic, noting that a certain degree of flexibility in those arrangements helps smaller carriers enter the market and increases their flexibility in obtaining operating agreements. At the same time, it ordered carriers providing those two arrangements to adhere to the ITU's CCITT Recommendation D.60 and stated that it would monitor such arrangements. Recommendation D.60 calls for 50/50 division of tolls between the terminal administrations with the cost of transit to be also divided equally between them. This arrangement generally results in a division close to 40-20-40, where the originating telephone company keeps 40% of the revenue, the intermediary keeps another 20%, and the remaining 40% goes to the carrier terminating the call.

7.4.2 Basic Voice

Until the early 1980s brought the advent of competition in international voice services, the problem of whipsawing had been confined to international record carriers (i.e., providers of telex service and other nonvoice services). AT&T was the only provider of voice services, and as such enjoyed equal bargaining power with foreign administrations.

As competition developed in the United States, AT&T's negotiating position began to erode. The issue arose for the first time in 1985 in a case before the Common Carrier Bureau [9] involving U.S.-Canada traffic [10]. However, the issue of whether the USP should be extended to voice services was not addressed there because the Common Carrier Bureau found that it should be considered in a broader rulemaking proceeding and not a lower level adjudication.

When the issue of extending the USP to voice services was finally addressed in 1986, the FCC found that the policy was created to address a set of market conditions rather than a particular technology or service type, and the USP was therefore explicitly extended to voice services. However, because of the particularities of the voice market, the FCC relaxed the waiver application process for voice carriers. This process will be discussed in Section 7.5.

Notwithstanding the relaxation, the FCC has denied requests for waiver of the ISP in certain cases. In 1990, AT&T and MCI introduced a new service that would permit callers to call the United States through an operator from any telephone in Spain. In return for access, the U.S. carriers would pay Telefónica de España a higher accounting rate. The carriers defended the arrangement by arguing that the service was a new and different offering, but the FCC ruled that it was merely another form of IMTS (or plain international voice service) and denied the waivers, finding that the carriers had been whipsawed [11].

7.4.3 Enhanced Services

The FCC concluded that it had the authority to apply the ISP to enhanced services. While it would not subject the providers of such services to the filing requirements imposed on basic service providers, it would continue to monitor the operating agreements for evidence of whipsawing and would intervene if necessary to prevent it.

In 1982, for example, NORDTEL, the Scandinavian PTT, announced it would take competitive bids from U.S. carriers to establish an operating agreement for the provision of enhanced services with a limited number of them. Several other European administrations followed suit. The FCC and the State Department intervened and pressured those administrations to drop their efforts to elicit competitive bids. The FCC formally declined to apply the USP to enhanced services in a later proceeding [12]. It concluded that its application to those services would not serve the public interest for three reasons. First, it would damage the United States' stand at the upcoming WATTC (ITU meeting) as a promoter of private-line services and unregulated and competitive enhanced services worldwide. Second, the authority granted to the FCC by the Communications Act, coupled with its ability to engage foreign administrations in bilateral negotiations, would permit it to address any whipsawing problems arising in enhanced services. Third, the FCC did not intend to strictly enforce the policy anyway. Enhanced services provided a lessened concern; because no entity entered the market as a dominant provider of those services, the opportunities for whipsawing were greatly reduced.

7.5 APPLICATION OF THE INTERNATIONAL SETTLEMENTS POLICY

The ISP is codified in Sections 43.51 [13] and 64.1001 [14] of the FCC's rules. Those sections direct carriers to inform the FCC and their competitors when they negotiate a special arrangement with a foreign carrier. Section 43.51 directs carriers to file with the FCC a copy of any agreement they enter into with a foreign carrier within 30 days of its execution.

The agreement must contain the accounting rate and the division of tolls. Then any carrier that subsequently enters into a new agreement or amends its agreement with a foreign administration on terms that are not identical to the equivalent terms and conditions offered to its competitors on the same route must so advise the FCC and file either a waiver request or a notification letter as appropriate. Carriers who are amending their operating agreements with foreign administrations are subject to the same requirements.

When a carrier is simply effecting a reduction in the accounting rate with another carrier, a notification letter containing the old rate, the new rate, and the effective date must be filed and a copy served on all the carriers providing similar service on that route. The new rate may become effective on the day the notification is made. If the notification merely reflects that a carrier is matching an accounting rate reduction obtained by a competitor, the reduction may be retroactive to the day the first carrier's reduction became effective. Any other modification of the terms of the agreement requires a formal waiver explaining the modification.

Waiver requests are subject to a 21-day pleading period during which any interested party may oppose the waiver or file comments concerning it. If no formal opposition is filed and the FCC staff has not found the waiver objectionable, it is automatically granted on the 22nd day without formal FCC staff action. The FCC will grant waivers when it finds that the changes in the terms of an operating agreement benefit consumers by encouraging new entrants and encouraging innovation and lower prices [15].

No single factor determines whether the FCC will grant a waiver. Rather, it will look at each request individually and consider all the factors present. Smaller carriers probably enjoy more leniency in the application of the ISP as a way to help them compete against the major players.

7.6 OTHER FILING REQUIREMENTS

Every nondominant international carrier operating an Earth station must file an annual employment report to document its compliance with U.S. Equal Employment Opportunity laws. In addition, it must file circuit additions reports for private lines, annual circuit status reports for all lines, and annual traffic and revenue reports [16]. A copy of every operating agreement and every agreement to interconnect private lines to the public switched network must also be filed [17]. The FCC compiles the data obtained from the reports and publishes them annually. A carrier that becomes affiliated with a foreign carrier must notify the FCC of that fact [18].

7.7 THE INTERNATIONAL ACCOUNTING RATES PROBLEM: A NEW FORM OF FOREIGN AID?

Today, because of competition in the United States, an outbound international call generally costs the caller significantly less than the same call in the other direction. The reason for this discrepancy is both simple and perverse. As we saw above, at least three telecommunications carriers are involved in any call to or from the United States: the local exchange carrier (for example, BellSouth) and the long-distance carrier (for example, AT&T) on the U.S. end, and the telephone company in the other country. In some countries, a fourth player, the foreign-end local exchange carrier, is also involved.

The carrier at the originating end of the call bills its customer for the call. The price the customer pays is known as the collection rate. The carrier must also compensate the receiving carrier for completing the call. That compensation between carriers is known as the accounting rate. As a result of lower calling prices caused by competition between carriers, the volume of calls originating from a few countries, such as the United States [19], has increased much more rapidly than that of calls originating from other countries. Because the originating carrier must compensate the receiving carrier, this situation has created a perverse trade imbalance to the detriment of U.S. carriers who must pay their foreign correspondents more than they receive from them. At the same time, because U.S. carriers must compete with one another for market share, they must outperform each other and offer competitive collection rates—a problem most of their foreign counterparts do not face.

The FCC estimates that the deficit increased from US$40 million in 1980 to more than US$2 billion in 1989 and US$2.9 billion in 1990, a 20.7% increase from the previous year [20]. If left unchecked, the deficit could reach US$7 billion by 1998 [21]. Another reason for this imbalance may be the off-peak or time-of-day discounts offered by U.S. telephone companies. A carrier must engineer its network to handle peak traffic. Outside business hours, it will have excess capacity that can be sold to its customers at a discount. Accounting rates, on the other hand, are not generally discounted in off-peak periods, or not discounted as significantly [22].

In 1990 the FCC turned its attention to accounting rates as a follow-up to its decision concerning the ISP. As part of its new ISP, the FCC decided it had to address above-cost international accounting rates. It noted, for example, that U.S. carriers were paying exorbitant accounting rates to countries in Asia and Europe that already have well-developed telecommunications infrastructures. Moreover, the FCC found evidence of discrimination by foreign administrations against U.S. carriers. Although intra-European accounting rates are not published, there was evidence that U.S. carriers were forced to pay higher accounting rates than their European counterparts to terminate U.S.-originated calls in several countries [23].

A number of foreign administrations have argued that high accounting rates are a necessary form of subsidy to help their governments develop their dated telecommunications infrastructures. However, the United States has maintained its position that accounting rates should be cost-based, and propounded this position at ITU CCITT meetings, and that lower, cost-based accounting rates are in the public interest. In 1992, the CCITT approved Recommendation D.140, which recommended that accounting rates reflect the cost of providing the service and nondiscrimination between operators. The FCC found that in the absence of progress by U.S. carriers in significantly lowering international accounting rates, it had the authority to establish certain regional benchmarks or rates that carriers may not exceed in their payments to their foreign counterparts. These benchmarks were established at $0.23 to $0.39 for Europe and $0.39 to $0.60 for Asia, and carriers were directed to continue to negotiate lower accounting rates [24].

7.8 GETTING AROUND THE ISP: INTERNATIONAL RESALE AND CALL-BACK

In the years since the ISP was adopted, carriers on both sides have sought ways to minimize its impact and avoid its effects. Two forms of ISP bypass have acquired particular notoriety in the past few years: private-line resale and call-back services.

7.8.1 The Issue of Resale

After addressing the first part of the accounting rates problem, the FCC turned to other aspects of its policy that could help encourage competition and exert downward pressure on accounting rates. In 1991, the FCC found that the benefits of unlimited resale in the domestic market could be extended to the international market as well and ordered all U.S. carriers to permit unlimited resale of their capacity, except for private lines interconnected with the public switched network [25].

Providers of private-line service typically charge by capacity rather than by minute of use. Therefore, permitting resale of private lines would allow foreign telephone companies to send traffic to the United States through a private line to be interconnected with the U.S. public network and thus evade the settlements process without granting U.S. carriers a reciprocal right to do the same. The FCC ordered U.S. carriers to permit resale of their capacity for the provision of public switched services (such as IMTS) and their private lines not interconnected with the public switched network. The ruling did not affect the ability of end users to interconnect private lines for their own exclusive use, or the ability of purely facilities-based carriers to offer interconnected private lines to their customers for their own private use. Private lines interconnected with the public switched network may only be resold to countries that afford "equivalent resale opportunities." To date, only the United Kingdom and Canada have been found to offer equivalent resale opportunities. (For more on resale of interconnected private lines, see Chapter 7.)

7.8.2 The Problem of International Call-Back

In response to the high collection rates of foreign administrations, some U.S. carriers have developed novel methods of providing service outside U.S. territory. Call-back generally consists of a switch located in the United States that is capable of identifying the number from which it is being called and of returning the call before it is answered. There are other forms of call-back as well, but the mechanism is similar in all. For example, an operator may answer the call for a few seconds and return it immediately with a U.S. line. Or a switch in the U.S. may constantly call a number in a foreign country. The ringer is hushed on the other end, but when the customer decides to place a call, he or she will immediately have a U.S. line available. This method is known as *hotlining*.

In the classical form of call-back, a customer in a foreign country, such as Zaire, calls a number in the United States. After three rings with no answer, the customer hangs up. Since no communication has been established, the caller is not charged by the Zairian telephone company. The U.S. switch then calls the caller back immediately and offers a U.S. dial tone to place calls on a U.S. line at U.S. prices. This arrangement circumvents the Zairian national carrier and causes it to lose revenues it would otherwise collect.

Call-back services constituted a market producing more than US$120 million in revenue in 1993. Affected governments and carriers soon began complaining about this situation and sought to have these services declared illegal. At the same time, in the United States, where call-back operators number in the hundreds, a number of companies that provided these services asked the FCC to rule that they are legal and advance the FCC's goal of increasing competition in the global market. Several countries, including Peru, Costa Rica, and Venezuela, declared call-back services illegal in their territory.

Call-back operators justify their services by declaring that they only circumvent the prohibitively expensive rates of PTTs that have used those rates for years as a hidden tax on their customers and traveling business people. In the Spring of 1994, the FCC declared call-back services legal. AT&T, which also loses revenue from certain forms of call-back (the more classical, incomplete signaling call kind), immediately filed a petition for reconsideration. Several foreign PTTs filed comments supporting the petition. In the summer of 1995, the FCC reaffirmed the legality of call-back services, but added that in the interest of international comity, carriers may not use their Section 214 authorizations to offer call-back service using uncompleted call signaling in countries where this service is explicitly prohibited [26]. Although the order limits the prohibition to call-back offered by means of uncompleted call signaling only, it is possible the FCC would extend its ruling to other forms of call-back if the question arose.

In the United States, international carriers are subject to certain rules designed to protect competition in the U.S. market and to encourage their foreign counterparts to reduce international accounting rates. These rules evolved because the U.S. telecommunications market began as a competitive market for certain services, while foreign markets were generally created as government-owned monopolies. The disparity between the number of players in the foreign market and in the U.S. market reduced the bargaining power of U.S. carriers, forcing them to give concessions to their foreign counterparts to obtain a competitive edge at home. The FCC intervened to correct this situation. As the market evolved, new issues arose. The FCC sees its role as protecting the U.S. consumer from abuse by promoting lower calling prices. In order to reduce calling prices, it has mandated resale of international lines and generally declared call-back services lawful. At the same time, to preserve the health and viability of the U.S. competitive market, it has imposed certain restrictions on what U.S. carriers may do.

Notes

[1] Communications Act, Section 201.

[2] Enhanced services are services in which a computer interacts with information provided over telephone lines (see Chapter 6).

[3] In fact, the service may be provided over the same facilities, as is the case in resale arrangements, where one carrier purchases bulk capacity from another at a discount and resells it to its customers for less than the underlying carrier would charge them.

[4] 2 FCC 592 (1936).

[5] As the FCC pointed out, "If the [Norwegian] administration should have a choice of two competing direct radio circuits, it is only natural to expect that it would favor that circuit from which it would derive the greater financial advantage." 2 FCC 598.

[6] 2 FCC 599.

[7] *Mackay Radio and Telegraph Co., Inc. v. Federal Communications Commission*, 97 F.2d 641 (D.C. Cir. 1938), citing *Texas & Pacific Railway Company v. Gulf, C. & S.F. Ry. Co.*, 270 U.S. 266.

[8] *In the Matter of Implementation and Scope of the Uniform Settlements Policy for Parallel International Communications Routes*, 59 RR 2d 982 (1086) CC Docket 85-204.

[9] The FCC may delegate its authority to rule on certain routine matters to the chief of the Common Carrier Bureau. 47 C.F.R. §0.291.

[10] *MCI Telecommunications Corp. et al.*, CC Mimeo No. 3874, released April 16, 1985.

[11] *American Telephone and Telegraph Co.; MCI Communications Corp.*, 67 RR 2d 1706 (1990).

[12] *In the Matter of Implementation and Scope of the Uniform Settlements Policy for Parallel International Communications Routes*, Further Reconsideration, 64 RR 2d 956 (1988).

[13] 47 C.F.R. §43.51.

[14] 47 C.F.R. §1001.

[15] See, for example, *American Telephone and Telegraph Co.*, 67 RR 2d 217 (Common Carrier Bureau, August 18, 1989).

[16] Section 43.61 of the FCC's rules, 47 C.F.R. §43.61, and new Section 43.82, 47 C.F.R. §43.82 (1995).

[17] 47 C.F.R. §43.51.

[18] 47 C.F.R. §63.11.

[19] Other examples that come to mind are Chile and the United Kingdom.

[20] A few examples of IMTS deficit are US$410,865,000 for Mexico, US$153,146,903 for the Federal Republic of Germany, and US$14,636,060 for Ireland in 1988.

[21] Appendix 3 to Further Notice of Proposed Rulemaking, *In the Matter of Regulation of International Accounting Rates,* CC Docket No. 90-337, Phase II, 6 FCC Rcd. (1991).

[22] P. Bernt and M. Weiss, *International Telecommunications*, Sams Publishing, 1993, p. 89.

[23] AT&T stated that while intra-European rates range from 0.33 SDR (US$0.47) to 0.55 SDR (US$0.78), American carriers must pay between US$0.50 and US$1.47 more to terminate their calls. 69 RR 2d 247.

[24] *In the Matter of Regulation of International Accounting Rates*, Phase II, 71 RR 2d 868 (1993).

[25] *In the Matter of Regulation of International Accounting Rates*, 70 Rad. Reg. 2d 156 (1991) as clarified in 71 Rad. Reg. 2d 862 (1992) and 7 FCC Rcd. 7927 (1993).

[26] *Via USA, Ltd.; Telegroup, Inc.; Discount Call International Co.*, File Nos. ITC-93-031, ITC-93-050, and ITC-93-054, Order on Reconsideration (Released June 15, 1995).

8 Privacy, Intellectual Property, and International Trade

The institutions and processes we have discussed so far are concerned specifically with the regulation of telecommunications, but other types of law and regulation also have a substantial impact on the movement of electronic communications and services across borders. These include regulation of international trade, the protection of copyrighted material, and the increasingly vexing problems of electronic privacy. We briefly consider each of these topics here.

8.1 INTERNATIONAL TRADE

One of the oldest uses of governmental power is the protection of domestic industries from foreign competition, and the use of diplomacy (and sometimes military power) to open foreign markets to those same industries. In the telecommunications industry, participation of carriers and manufacturers in foreign markets was frustrated, historically, by the almost universal practice of licensing telecommunications carriers and protecting their monopolies against both foreign *and* domestic competition. Under the traditional system, if telecommunications companies did sell goods or services in foreign markets, they probably did so only as vendors to the PTTs; the question of foreign suppliers selling equipment or providing services in competition with the PTTs simply did not arise.

When some countries began to permit competition in their domestic telecommunications markets, however, international trade in telecommunications became an issue. The most important such development was the opening of the AT&T network to competition in the United States, with the resultant opportunities for foreign vendors to compete with AT&T and other U.S. companies. With the markets of other countries still closed to competition, U.S. companies—many of them large players—began to experience a substantial trade deficit in telecommunications markets. The U.S. government heard the complaints of these companies, as well as similar complaints from the banking and data processing industries, and in the mid-1970s the United States became the leading advocate of reduction of trade barriers in services generally, and in the telecommunications industry in particular.

For a number of reasons, the United States chose to focus its liberalization effort on the General Agreement on Tariffs and Trade (GATT)—a complex set of rules and agreements by which many nations have reduced trade barriers between each other [1]. The GATT's advantages as a vehicle of change included the economic influence of its signatories, the GATT's commitment to free trade (lacking at the time in other international organizations, such as the ITU), and the GATT's effective enforcement mechanisms. Incorporation of telecommunications into the GATT framework would not only help to open the markets of GATT signatories, but would create a powerful precedent for agreements with other nations as well [2].

The United States faced two large obstacles, however, in its advocacy of this position. The obvious problem was the resistance of many foreign PTTs to competition; the other was the fact that most telecommunications markets were *service*, rather than *product,* markets. Throughout history, it seems, goods flowed across borders, but services rarely did. So international trade agreements (including those of the GATT) involved only tangible things, and no one knew how to quantify and regulate trade flows involving services.

In spite of these obstacles, the U.S. Congress decided as early as 1974 that the Administration should work toward the inclusion of services in the GATT. The Administration shouldered the burden, but progress was slow. The United States was unable to make much headway in the Tokyo Round of GATT negotiations in the 1970s, but eventually persuaded the Organization for Economic Cooperation and Development (OECD) to study the subject, and in 1982 persuaded the GATT signatories to sanction similar studies by member countries. Finally, in 1986, after the United States had suggested that it might pursue bilateral solutions, the subject of trade in services was put on the GATT agenda.

In 1989, at the Uruguay Round Mid-Term Review meeting in Montreal, the GATT signatories at last agreed to a set of principles to govern the inclusion of services in the GATT. It was decided that negotiations concerning trade in services would proceed on a separate track from negotiations concerning trade in goods, and that service negotiations would be further divided into talks on particular sectors—including telecommunications. Any agreements reached in the sector discussions would be formalized in documents called *annexes.*

Discussions concerning trade in telecommunications services eventually produced a Telecommunications Annex, which recognizes the following principles:

1. Information concerning terms and conditions of service and network access must be publicly available.
2. Access to public networks must be available on reasonable and nondiscriminatory terms.

3. Pricing of public network services must be based on cost.
4. Connection of private networks and customer-provided equipment to the network must be permitted.
5. Any conditions imposed on access to the public network must be intended to protect the technical integrity of the network and the ability of the public network provider to meet its service obligations.

In the view of some potential competitors and large communications users, the annex offers theoretical liberalization, but permits the PTTs, as a practical matter, to impose substantial restrictions in the name of protecting their universal service obligations [3]. In the meantime, sector talks continue on the question of opening up basic long-distance services to competition in GATT signatory countries.

While this brief discussion has concentrated on the GATT as an instrument of multilateral trade liberalization in telecommunications markets, other organizations and agreements also are important to this process. The North American Free Trade Agreement (NAFTA), for example, takes an approach to telecommunications that is largely based on the GATT Telecommunications Annex.

Finally, anyone wishing to understand the global picture concerning trade in telecommunications should become familiar not only with the many bilateral agreements that have been concluded among nations, but also with the work of groups like the OECD and the Asia Pacific Economic Cooperation group as well.

8.2 ELECTRONIC PRIVACY

Informational privacy has been defined as "the claim of individuals, groups or institutions to determine for themselves when, how and to what extent information about them is communicated to others" [4]. The ability of individuals to enforce this claim is far more problematic today than it was in simpler times.

Before the invention of electronic communications, people typically protected their personal information without the assistance of law or technology: a closed door ensured that a conversation was not overheard; the professional discretion of clergy, physicians, bankers, and lawyers kept financial and personal affairs confidential; and the information-processing capacity of paper and ink limited the accumulation, transfer, and use of data about the lives of private citizens. Informational privacy was largely a matter of self-help, supplemented by force of law only when needed to discourage the acquisition of information by theft, coercion, or in the course of abusive governmental searches and seizures.

The advent of electronic technologies made it far more difficult for individuals to control the acquisition, disclosure, and use of information about themselves. The telephone and telegraph carried personal messages by wire over great distances, making the closed door ineffective and exposing the user of these technologies to electronic eavesdropping at any point along the circuit. The invention of the digital computer, which stored and manipulated data at orders of speed and quantity unheard of in the age of paper and ink, raise the threat of unauthorized access to and disclosure and use of personal information stored in—and capable of instantaneous transfer among—databases maintained by governmental agencies and commercial enterprises.

These threats to informational privacy have been multiplied by the growing popularity of the personal computer and the ubiquitous interconnection of computers over telephone lines. With the growth of electronic mail, bulletin boards, electronic funds transfer, and remote computing services of all kinds, we are conducting more and more of our private conversations and business transactions over insecure telephone lines and feeding more and more information (knowingly or unknowingly) to public and private networks and databases. As computing and telephony continue to converge, the range of situations in which we do not control the acquisition, disclosure, and use of information about ourselves will increase.

Most of the developed countries have recognized the threat to individual privacy posed by the transfer and storage of computerized data, and have passed legislation that offers some protection to individuals. The scope of protection and the mechanisms by which it is enforced, however, vary widely.

In the United States, for example, the Privacy Act of 1974 imposes substantial restrictions on the ability of the federal government to accumulate and use information about individuals, and confers on individuals significant rights to be informed of, examine, and correct personal information stored in government databases. U.S. laws do not, however, offer any similarly comprehensive controls on the accumulation and use of data by commercial enterprises, and even the Privacy Act's protections rely heavily on individual initiative rather than government regulation for their enforcement.

Many other advanced nations have enacted privacy laws that go well beyond those of the United States and that impose restrictions equally on private and governmental databases [5]. The Swedish Data Act, for example, requires all databases containing information about individuals to be licensed and gives a Data Inspection Board broad powers to impose conditions on individual licensees. Germany forbids data collection except as expressly permitted by its statutes and regulations or by consent of the subject of the collected information, and has a federal commissioner for data protection (BfD) who regulates data collection by the federal government and all private databases. (The individual German states regulate the data protection practices of their own administrations.) In the United Kingdom, the Data Protection Act requires registration of all databases used to store personal information, and database operators who exceed the terms of their registrations or otherwise violate the Act are subject to cancellation of their registrations and other penalties.

From the standpoint of international telecommunications, the most important development in the law of data protection is the European Union Data Protection Directive, adopted by the EU at Cambridge, England, in September of 1995. This directive, which was first proposed in 1990, requires all member countries to implement a minimum set of data protection practices. The standards include notification of persons on whom data is collected, limitations on release or use of data for purposes other than those disclosed to the subject, and a right of individuals to inspect and correct any personal information contained in a database. EU members were given three years to bring their laws into line with the directive.

Besides the impact of the directive on member nations (like Italy) that had not yet enacted privacy legislation, the most significant effect of the new law may be on the electronic movement of data among countries. Even before the directive became the law of the EU, some European countries had declined to disclose database information to U.S. authorities because the U.S. privacy laws do not protect information (particularly private-sector information) as rigorously as European statutes require. The new directive actually forbids the export of data to countries found by the EU authorities to have inadequate privacy laws—a provision that could lead to restrictions on the electronic transfer of data to the United States [6].

8.3 INTERNATIONAL COPYRIGHT PROTECTION

As electronic technologies increase the ways in which information is stored and conveyed, and as those technologies accelerate the dispersal of information around the world, the problems of international copyright protection become more urgent. Those problems have been with us for a long time and derive from the differing approaches to copyright protection taken by different countries and cultures. To put it as briefly as possible, the very notion of copyright protection is a largely Western invention, and even in the Western world the concept has different meanings that lead to conflict when the works of authors move across borders. This section considers, in a very brief outline, the background of some of these disputes and some recent efforts to harmonize different approaches to copyright protection.

8.3.1 The Berne Convention and Copyright Protection in the West

The incentive of any country to protect the works of foreign authors depends strongly on whether that country is a net importer or exporter of literary and artistic works. Net exporters will demand protection for their authors, while net importers find that the benefits to their domestic consumers from cheap, pirated editions outweigh the pain caused to their domestic authors by foreign pirating [7]. Until the late nineteenth century, a rough calculus of this kind determined the attitude of governments toward the works of foreign authors: net importers (like the United States) denied protection, while net exporters (like France, which had at the time a far richer literary tradition) took the opposite view.

Of the early efforts to establish multilateral copyright protection based on neutral principles, the most notable was the 1886 Berne Convention for the Protection of Literary and Artistic Works (the Berne Convention)—an agreement that, with much modification, remains in force today. As first ratified among some of the principal European nations, the Berne Convention provided for protection of works on the basis of so-called *national treatment*—that is, the principle that each signatory would extend to the works of authors from other treaty members the same copyright protection that it offered to its own nationals [8].

The Berne Convention has not, however, prevented significant differences of copyright treatment from arising, even among its signatories. Those differences are embedded in the different legal cultures of the signatories and have been aggravated by new technologies.

The primary difference among Western nations with copyright traditions is the distinction between *author's rights* countries and the more pragmatic culture typified by the United States. Put very crudely, copyright protection in the United States extends to any fixed, literary, or artistic expression, where such protection seems necessary to encourage literary or artistic output, while in France and other European jurisdictions, copyright protection centers on authors, rather than business enterprises that reproduce and distribute authors' works, and on creative works, rather than merely mechanical reproductions of those works. Based on these principles, Americans have little trouble extending copyright protection to photographs, sound recordings, films, and the companies that produce them, while author's rights countries find these devices an uneasy fit with their tradition.

European countries nonetheless have made the conceptual leap needed to protect mechanical works and their producers where no protectionist impulse pulled in the other direction. (Photographs and motion pictures were protected by characterizing the photographers and cinematographers as authors.) With sound and motion picture recordings, however, European signatories to Berne took a different view. France, for example, collects a royalty on each blank audio or videotape and distributes part of the royalties to its domestic performers and producers of recorded performances, as well as to the authors of the works that have been performed and recorded. France also distributes royalties to U.S. authors, but not to U.S. performers and producers. The reason is that the performers and producers are not authors under French copyright law and therefore have at most so-called *neighboring rights*, which a Berne signatory is not required to honor under the terms of the convention [9].

These and other protectionist efforts have caused some acrimony among the Western nations and have led to new efforts to harmonize copyright law in a way that takes new technologies into account. One such effort, for example, was the EU's copyright directive for protection of computer software, which nearly collapsed in a controversy over the right to reverse-engineer software programs. Even more difficult challenges lie ahead as the EU takes on protection of computer databases and an effort to clarify the rights of authors.

8.3.2 Recognition of Copyright in Asia and Elsewhere

As we have noted, the idea of copyright protection is a largely Western invention and is even in the West observed best by those countries that have something to gain by doing so. In the rest of the world, which tends to be a net importer of information, respect for Western notions of copyright has been nearly nonexistent [10].

The greatest problem for Western authors and producers was Asia, which developed an enormous pirate industry for everything from motion pictures to medical textbooks. Fortunately, in 1990 the People's Republic of China adopted a copyright law containing a number of Berne principles, and the rest of Asia (particularly the countries with strong trading ties to the West) are under pressure to follow suit. At the same time, the Chinese have shown only an intermittent inclination to crack down on their lucrative pirating industry, and enforcement of Asian statutes and international undertakings with Asian countries is likely to be an ongoing problem.

Notes

[1] As Chapter 12 explains, the GATT began in Western Europe in 1948 as a trade agreement covering basic industrial commodities and manufactured goods. The GATT has since extended beyond Europe into other markets, but still does not apply with full force to agriculture and textile goods.

[2] The U.S. had alternatives, of course, to the GATT. One possibility was simply to wait for foreign administrations to open their markets in response to internal pressures—a process that acquired great momentum in the 1980s, but seemed not to be happening at all in the 1970s. Another option was to pursue bilateral agreements with individual nations—a course the U.S. has followed in a number of cases, but in parallel with, rather than as a substitute for, its GATT negotiations.

[3] See, for example, *U.S. Telecommunications Services in European Markets*, U.S. Congress, Office of Technology Assessment, 1993, p. 148.

[4] A. F. Westin, *Privacy and Freedom*, 1967, p. 7.

[5] In fact, of the major industrial nations, only Canada, the U.S., and Australia limit the scope of their data protection laws to government.

[6] European data protection experts have raised this issue with U.S. authorities many times over the years, notably at data protection conferences of the OECD.

[7] In the absence of treaty provisions to the contrary, the rights of authors in foreign countries are determined by the laws of the foreign country.

[8] This is in contrast to the common, alternative principle of reciprocity, according to which country A will extend to country B's authors the same treatment that country B extends to country A's authors. The U.S., incidentally, did not formally agree to be bound by the Berne Convention until 1989, although it did adhere to its own creation—the Universal Copyright Convention—which was signed by many of the Berne signatories and was also based on national treatment.

[9] The Berne Convention requires each signatory to apply the principle of national treatment to protect *works* and *authors*, but each Berne signatory may interpret those terms according to its own legal tradition. The rights of nonauthors, or rights of nonworks, may be classified by the signatory as neighboring rights, which it may treat as inferior to the intellectual property rights it is bound to respect under the principle of national treatment.

[10] In 1967, a number of developing nations at a Berne Convention meeting in Stockholm demanded to be relieved from the full discipline of a treaty to which they became signatories when under colonial domination. While this problem was compromised in an agreement reached in Paris in 1971, the incident demonstrated the hostility of non-Western nations to a copyright regime in which they perceive little benefit to themselves.

Part II: Telecommunications Within Nations

Domestic Telecommunications Regulation: A Global Survey

9

If this book had been written 20 years ago, the following chapters, which describe the state of domestic telecommunications regulation within specific nations and regions of the world, would present a static picture. Almost without exception, telephone service was provided by public or private monopolies. Where the service provider was a government agency, service was funded through a combination of user charges and transfers from the public treasury. Where the service provider was a private monopoly, service was funded by user charges set—or closely scrutinized by—a government agency. Whatever the details of individual systems, decisions concerning pricing, investment, and terms of telephone service were ultimately under political control.

Today, however, the picture is of regulatory systems at various stages of an inexorable march toward open markets and competition. While each of the countries we shall survey has followed its own course through this transition, the process displays enough common themes to support some introductory generalizations. This chapter provides some of this background.

9.1 THE RISE AND FALL OF MONOPOLY TELECOMMUNICATIONS

Telegraph and telephone services were first widely deployed in the nineteenth and early twentieth centuries, chiefly in the industrial societies of Western Europe and North America. In both regions, the new services quickly came under the control of monopolies: the governmental postal monopolies (expanded to become PTTs) in Western Europe and private monopolies in the United States and Canada.

Over the decades, these monopolies became associated with strong political and economic interests. The European PTTs, in particular, were allied with the press and the publishing industry (which enjoyed discounted postal and telegraph rates) and with organized labor. And in countries with a telecommunications equipment manufacturing sector, the monopoly carriers became important as captive markets for that industry's products.

The telecommunications monopolies and their political patrons developed various policy justifications for their continued insulation from competition. One such rationale was the argument that the enormous investment required to build a wireline telephone network made telecommunications service a natural monopoly [1]. Another was the argument that competition would destroy the monopolies' internally subsidized rate structures and cause increases in basic, residential, and rural service. Another concern was that a unified telephone and telegraph network, under government control, was essential to national defense. A fourth rationale, especially popular in France in the 1970s and 1980s, viewed the unified telephone network as a tool of international economic competition [2].

Beginning at different times in different countries, the monopolies began to lose some of their economic and political power. The new pressures came from a number of directions. On the economic side, business interests with considerable clout of their own became dissatisfied with the monopoly environment, either because they wanted to provide competing products and services or because they wanted to purchase products and services that the monopolies did not offer. So, for example, equipment suppliers wanted to sell PBXs and telephones, data processing companies wanted to sell on-line computer services, microwave radio operators wanted to offer long-distance service between major business centers, and large users of telecommunications, including the manufacturing and financial sectors, wanted to buy the new products and services [3]. On the policy side, the arguments for monopoly were eroded by technology and shifts in public and bureaucratic attitudes. New technologies, such as microwave and mobile radio, suggested to regulators that some telephone service markets were no longer natural monopolies. And growing distrust of large corporate and governmental institutions—particularly in the United States—made the public less sympathetic to the idea that a monopoly system would protect the interests of ordinary customers better than a system of open competition [4].

These changes made themselves felt with dramatic force in the United States, where most telecommunications products and services were provided by a single corporation of over one million employees. As we discuss more fully in the chapter on domestic telecommunications regulation in North America, the U.S. Department of Justice sued AT&T under the antitrust laws, effectively forcing the company to divest all of its local telephone operations from its long-distance, research, and manufacturing operations. Subsequently, a federal court and the FCC have enforced elaborate rules intended to ensure that AT&T and its former operating companies treat their competitors fairly.

Other countries have followed their own paths toward competition in domestic telecommunications. These efforts have typically included some or all of three elements: the privatization of government-owned PTTs, control of new entry into telecommunications markets to protect the viability of the PTT, and the promulgation of rules to control the lingering market power of the former monopolists. The following sections discuss each of these processes in some detail.

9.2 RESTRUCTURING AND PRIVATIZING THE PTTs

Where governments have decided that their PTTs are a drain on the treasury or lack the capital and incentives needed to provide up-to-date services, they may choose from a number of possible solutions. One possibility is to keep the PTT in public hands, but take various measures to commercialize it—that is, improve its funding and efficiency. Efficiency may be achieved by eliminating taxpayer subsidies and requiring the PTT to sustain itself entirely through user charges. Funding may be improved by permitting the PTT to enter into joint ventures that let other (often foreign) companies share in the profits of the enterprise in exchange for contributing their money and expertise.

While these approaches have their merits, they typically do not match the array of benefits that sale of the PTT to private investors can provide. When privatization works as advertised, it can bring in quantities of capital not available from the public treasury, bring to telecommunications markets the efficiency and innovation associated with the profit motive, and free management from political constraints. Privatization removes the temptation to use revenues from telecommunications service to offset the deficits of the postal service or other public programs, instead of for reinvestment in telecommunications service and facilities. For developing countries, sales of shares in the privatized companies may vitalize domestic capital markets and bring foreign capital into the economy of the privatizing country.

In order for a privatization to realize its aims, however, the government must manage the process carefully before, during, and after the sale. Before the sale, the government must do at least two things: first, it must reorganize the PTT as an independent, profit-seeking entity (a process often called *corporatization*); second, it must put in place—or at least articulate—the system of regulation under which the privatized entity will operate. Each of these processes is essential if prospective investors are to feel comfortable with the decision to place their money with the enterprise.

The first step, corporatization, typically involves two processes that we might call the separation phase and the business organization phase. By the separation phase we mean simply the isolation of the PTT's telecommunications operations from postal or other functions for which it may have been responsible [5]. (It also may include the separation of the PTT's regulatory functions, which will remain with the government, from the business operations that will be privatized and regulated.) By the business organization phase we mean (1) the introduction of separate capitalization and a system of accounting by which the financial performance of the privatized entity can be measured, and (2) the restructuring of the entity to make it more efficient and profitable in a nonmonopoly environment [6].

The second preprivatization step—defining the system of regulation under which the entity will operate—requires that at least the broad outlines of the new regime be identified. Will the new entity be protected from competition for some period of time? If so, will protection extend only to its basic services, or will it have a monopoly of the more profitable, premium services? Will the company's earnings be restricted, or will it enjoy the profit potential of incentive regulation [7]? For potential investors, the answers to these questions may be as important as the results of the corporatization process.

The pressure on the government to complete the corporatization process and put a detailed regulatory system in place before privatizing will vary, depending on the kind of investor the government is seeking. If the investors will be experienced telecommunications companies who plan to run the entity, those investors may be content with a free hand to complete the task of corporatizing the entity themselves, and may negotiate the terms of regulation with the government as part of the concession under which they will operate. Purely financial investors, who will not operate the entity, are more likely to look for a fully corporatized entity within a well-defined regulatory environment. (Both types of investors, of course, will be concerned with the country's political environment and the ongoing ability of the government to honor its commitments.) Once the government has made the PTT and its environment attractive to investors, it must decide how the sale will be handled. The government is confronted at this stage with a number of alternatives [8].

One approach is to sell shares in the PTT on the domestic (and perhaps foreign) securities markets. Such offerings are complex and expensive undertakings. As noted earlier, they require extensive prior corporatization of the PTT. They also require compliance with the securities laws of each country in which the offering will be placed, and involve substantial fees to brokers and consultants. Where these obstacles are overcome, however, public offerings give the government access to diverse sources of capital (especially institutional investors) and pave the way for the company to return to the public capital markets in the future.

Privatization may also be accomplished through a negotiated sale with a single buyer or a syndicate of buyers. This method, if well executed, has the potential to generate high proceeds with minimal transaction costs, particularly if the field of potential buyers is narrowed through an auction process. Negotiated sales are especially appropriate where the government wishes to sell the PTT to a foreign telecommunications company (or companies) that will operate the privatized entity. Where the selling country has a large unmet demand for telecommunications services, the foreign operator may see that country as a field for rapid growth that is no longer available in the operator's home market, and may be willing to pay a premium for that opportunity.

Another privatization method, especially attractive to governments with a heavy burden of debt to foreign banks and an inability to attract conventional equity investment, is an exchange of debt owed by the government for a debt or equity position to be held by the creditor bank in the privatized entity. These arrangements give the lender banks the incentive to become involved in the management of the privatized carrier and to use their prestige to bring in additional investors with the expertise needed to develop and operate the company. For banks that are overexposed to a governmental borrower, conversion of the debt to an interest in a privatized company may be an opportunity to build a healthy company that can repay the debt and—if all goes well—return additional profit to the bank and its partners as well.

In the real world, variants of these methods tend to be used. For example, a public offering may not be made all at once, but may be consummated in a series of offerings; or the government may make a negotiated sale, not of the entire PTT, but of a partial interest. (Such a negotiated sale may be combined with a public offering of the remaining equity.)

Whatever the method used, all privatizations raise some common issues. One of these issues is the extent of foreign investment, which many governments wish to limit. Another issue is the ongoing stake of the government, which may wish to retain control as an investor with special, statutory rights. Concessions to national feeling and governmental sensitivity in these areas may be necessary in order to secure political support for the privatization.

After the sale, the government must regulate in a way that achieves both profitable operation and public benefits. If the goal is to achieve the benefits of competition, then the economic power of the privatized entity must be restrained at the same time that it is allowed to compete and earn a reasonable return on investment. We say more about these problems in the next section.

9.3 REGULATING THE TELECOMMUNICATIONS SECTOR

Most economists will tell us that regulation of an industry is a poor substitute for competition, and that where competition is feasible the regulator should step aside. Yet when governments introduce competition into their telephone industries, they often put in place systems of regulation that limit the markets in which competition will operate and dictate the decisions the dominant carrier may make concerning pricing, service offerings, and interconnection of its facilities with those of other carriers. And they do this for reasons that even an economist cannot dismiss out of hand.

For example, limitations on the markets in which competition will be allowed are usually intended to preserve the internally subsidized rate structures by which the dominant carrier keeps ordinary, local voice telephone service priced below the cost of providing the service. To avoid politically unpopular increases in basic service rates, governments may identify the services that are providing the subsidies to basic service and grant the dominant carrier a continuing monopoly over those services. These restrictions, along with licensing requirements and other controls that limit the markets in which competition will occur and the number of participants in those markets, are often referred to as *entry* regulation.

Similarly, limitations on the business behavior of the dominant carrier are based on a real concern—specifically, that the local exchange telephone network is (and for the foreseeable future will continue to be) a monopoly, with potential for abuse of customers and competitors alike. Customers can be abused through shoddy service and excessive or discriminatory rates; and providers of long-distance, value-added, cellular, and other services, who cannot reach *their* customers without access to the local network, may be subjected to discriminatory terms of access to the dominant carrier's facilities. Regulatory control over these practices is exercised through rate regulation, access regulation, and quality-of-service regulation.

Each of these kinds of regulation has taken different forms in different countries. The following discusses some of the principal approaches and also describes some of the institutional arrangements through which regulation is exercised.

9.3.1 Regulating Entry

The core business of any PTT is the provision of local and long-distance telephone service to residential and business customers. In setting their rates for these services, the PTTs historically have charged a premium over cost for long-distance, business, and urban services, and have charged less than the cost of service for local, residential, and rural services [9]. The original purpose of these pricing policies was to encourage extension of service to low-income customers and remote communities in the belief that expansion of the network increased its value to all users; but of course the subsidized rates, quite apart from their official rationale, hardened into entitlements with their own political constituencies [10].

This system of internal subsidies can work well so long as all telephone services are provided by a monopolist. With no alternative vendors available, business and urban customers, along with heavy users of long-distance service, have no choice but to pay rates that subsidize other users of the network.

If the government decides to permit unlimited competition in telephone services, however, the subsidies are jeopardized. New carriers may lack the resources or desire to duplicate the local exchange networks, but they can build long-distance facilities relatively cheaply using microwave radio or fiber-optic technology. Without the embedded cost of local network facilities to recover and without any local services to subsidize, the new carriers can charge rates for long distance that reflect only the cost of providing that service. Long-distance customers (especially businesses that rely heavily on toll calling) can be expected to rush to the new service, and the PTT [11] will lose the revenues it needs to keep local rates artificially low. As a result, local rates must rise until they are high enough to recover the cost of local service. If customers are unwilling or unable to pay those rates, then the government must either provide the PTT with support from the public treasury or permit the PTT to abandon local service to those who cannot pay.

This, of course, is precisely the scenario all governments wish to avoid [12], and their usual response to this problem is to limit entry into some or all of the markets that subsidize, or might be used to subsidize, basic service. This is typically accomplished by creating a category of *reserved* services that includes both the subsidized services and enough subsidizing services to support basic rates [13].

One such approach declares the entire voice telephone market a reserved service. Where this restriction is imposed, all local and long-distance voice services, both business and residential, continue to be provided exclusively by the dominant carrier. This leaves the PTT free to continue to charge a premium for voice service provided to businesses, urban customers, and users of long distance, and to apply those revenues to the support of local service for residential and rural customers.

Another approach, taken until recently in the United States and still followed in Japan, is to reserve all local voice service to the dominant carrier but permit competition in long distance [14]. This approach has the virtue of introducing market incentives to a service that is especially important to the business community, but it may also create downward pressure on rates for a service that is a prime source of subsidies for rural and residential customers. For this reason, countries that permit long-distance service competition have tended to introduce such competition gradually (as in the United States) or to limit the number of long-distance licensees (as is done in Japan and, until recently, in the United Kingdom).

Many countries that reserve some class of basic services to the dominant carrier also recognize the needs of large users for private networks, access to online data processing, and other services that the dominant carrier may not provide or may not provide at an acceptable level of economy or technical sophistication. So, for example, business customers with a high volume of voice or data traffic may wish to lease transmission facilities from the established carrier and use those lines for point-to-point or switched transmissions among their own business locations. Other entities, perhaps lacking the traffic volumes to justify leasing lines on their own, may wish to share facilities leased from the dominant carrier with other users; while still other businesses may find it profitable to lease lines from the dominant carrier and resell capacity on those lines to others on a retail basis [15].

Each of these uses of leased lines presents some threat to the revenues of the dominant carrier. When a commercial concern leases lines and resells them to third parties as part of a telecommunications service, it may compete directly with a service offered by the dominant carrier. When an end user leases lines for its internal use, it deprives the dominant carrier of the more profitable opportunity to provide that end user with a complete transmission service. For this reason, a number of governments still prohibit the leasing of telephone lines, and many governments that do permit leased lines forbid resale or impose other restrictions on the uses to which customers may put them. (So, for example, in some countries it is permissible to use leased lines, but not to connect those lines with the public switched telephone network; and in other countries, it is permissible to provide switched transmission of data, but not voice, to third parties over leased lines.)

One of the more common methods of controlling the extent of leased-line competition is to permit leased lines to be resold only in connection with value-added services. Some explanation of this restriction is in order.

A value-added service includes, but in some sense goes beyond, the electronic transmission of voice or data communications. It stores, forwards, performs data processing operations upon, changes the format of, or otherwise manipulates transmitted information. Common examples of such services include remote data processing, conversion of data transmissions between different data communication protocols, packet switching, electronic data interchange, automatic ticket reservation, conference calling, voice messaging, automatic credit card verification, and e-mail.

Governments that permit competition in value-added services but not in ordinary telephone service face stringent problems of definition and enforcement. If the definition of value-added service is too lenient, entrepreneurs will resell basic transmission services by adding some trivial enhancement that is said to add value to the lines leased from the dominant carrier [16]. If the line is drawn too narrowly, innovation will be stifled in markets that pose no threat to the viability of the established carrier.

At the same time, of course, even the most sensible definition may be stretched beyond recognition where there is a political motive to expand or contract the class of subsidizing services. Some European Community countries, for example, have argued that packet-switching networks and fax services should be reserved, even though those services clearly go beyond mere transmission and are value-added services under any reasonable definition.

Finally, we should be aware that the distinction between basic and value-added services is used for different purposes in different regulatory systems. Notably, the FCC in the United States has adopted elaborate rules based on a distinction between basic and enhanced services [17]; but the purpose of those rules is to ensure that when dominant carriers provide enhanced services in competition with other providers, they do not use their control over the local network to impose technical and pricing disadvantages on their competitors. Their purpose is not—as with the value-added service licensing rules of other countries—to protect the dominant carriers from competition in the provision of basic services [18].

The categories of reserved services we have discussed—whether defined to include all voice service, only local voice service, or all basic voice and data transmission service—leave a number of important services open to competition. Besides the value-added services already mentioned, the privatized PTTs generally do not have monopolies of telephone equipment, inside wiring, and mobile telephone services, and they may also face competition in international service, satellite Earth stations, and satellite transmission services. Competition in these nonreserved markets may not be unrestrained, however. Governments commonly use their licensing authority to control the pace of new entry and the number of competitors in these markets [19].

The final frontier of competition—and the last redoubt of reserved service—is the local exchange network. Only a handful of countries—the United Kingdom, Finland, and the United States among them—have opened the local loop to competition, and those efforts have met with mixed success [20].

One problem posed by local competition is the new entrants' need to resell and interconnect with the local facilities of the dominant carrier. If new entrants had to build facilities connecting all of their local customers with each other, duplicating a substantial part of the existing wireline exchange, or if the new entrants' customers could not reach customers of the dominant local carrier, then competition would be unfeasible. Regulatory regimes that mandate local competition, therefore, must also mandate reasonable terms of resale and interconnection.

Another problem posed by local competition is the familiar threat of lost support for rural and residential service. In addition to the internal subsidies by which long distance supports local, residential, and rural service, the rate structures of monopoly carriers also include subsidies among the local rates charged to different classes of customers. Specifically, urban customers tend to pay a premium over cost for their service so that rural customers can have affordable local service, and business customers pay substantially more for essentially identical services than residential customers. If new local service providers are permitted to target these urban and business customers and lure them away with cost-based rates, then the local rates of other customers must rise or regulators must find some way to replace the revenue support lost to the new carriers [21].

Still another obstacle to viable local competition involves the use of telephone numbers. Specifically, customers are less likely to take service from a new carrier if they must give up their old telephone number to do so and if they must dial a longer string of digits to place a call through the new carrier. Solving these problems—referred to respectively as the *number portability* and *dialing parity* issues—requires reprogramming of local carrier switches so that they will associate callers' numbers with the local carrier to which the call must be routed, and they also require, in the view of many, the appointment of neutral administrators to assign telephone numbers and conduct long-range planning for the numbering system. Only regulators who are prepared to mandate these steps can ensure a competitive local service market.

In spite of these complexities, open competition in all services—including local exchange service—is likely to become the norm rather than the exception.

9.3.2 Regulating Price

When all telecommunications services are provided by a government agency, pricing of those services is a political decision. Like rates for government mail delivery and government railway service, the PTT's telephone rates represent a complex accommodation of the demands of users for cheap service, the willingness of some customers to subsidize other, politically favored customers, and the willingness of taxpayers to make up aggregate revenue shortfalls with transfers from the public treasury. The function normally served by pricing in a market economy—as a method of rationing goods and services by charging consumers the full cost of providing them—plays at best a secondary role in the public PTT environment.

When the PTT is commercialized or privatized and is required to support its operations from the rates it charges its customers, pricing becomes a means of maximizing the PTT's profits. If the PTT retains a monopoly over some or all telecommunications markets and no system of rate regulation is imposed, pricing probably will be used to secure monopoly profits through excessive rates and discrimination among customers.

The traditional solution to this problem is to institute a system of earnings, or rate-of-return, regulation. Under this system, the carrier is permitted to set its rates at a level that will return revenues, at anticipated levels of demand, sufficient to cover the carrier's costs plus a reasonable return on investment. So long as the carrier's overall earnings do not exceed the allowed rate of return, rates for individual services are scrutinized only when they are in some sense unreasonable or "unreasonably discriminatory."

When the government has the resources and authority to determine the carrier's costs and enforce compliance with legal limits on earnings, rate-of-return regulation does a reasonable job of preventing carriers from earning monopoly profits. In recent years, however, some governments have concluded that rate-of-return regulation, by effectively guaranteeing recovery of all costs incurred by the carrier, also inflates rates by encouraging overinvestment [22]. In order to curb this tendency and replicate the cost-cutting incentives of competition, some regulators have adopted systems of incentive, or price cap, regulation.

The incentive approach was first adopted in the United Kingdom, and that country's system is representative of the method. Under the U.K. system, the rates for regulated services of British Telecom (the only company subject to rate regulation in Britain) are classified into *baskets*, and the Office of Telecommunications (OFTEL) then sets an upper limit, or cap, on the weighted average of rates in each basket [23]. At the end of some review period (usually four or five years), the price cap is reduced to match the anticipated average efficiency gains in the industry offset by anticipated inflation [24].

The price cap formula is designed so that by achieving the industry average of efficiency gains, British Telecom should be able to reduce its rates by the required amount and still make a normal rate of return. But if British Telecom can achieve greater cost reductions than the industry average, it will be allowed to keep the resulting higher return on investment. In other words, British Telecom, like any corporation in conditions of competition, earns higher profits when it achieves above average efficiencies.

While price cap regimes are potentially quite effective at making carriers more efficient, they are less effective at preventing—and in fact may be said to encourage—supracompetitive profits. For this reason, some incentive plans (e.g., the one implemented by the FCC to regulate the rates charged by AT&T and the larger local exchange carriers for interstate services) require that when earnings exceed a certain percentage, part of the carrier's profits must be shared with ratepayers in the form of refunds or rate reductions [25].

While rate and earnings regulations may be sufficient responses to the ratemaking problems posed by a pure monopoly environment, the problems posed when the dominant carrier offers both competitive and reserved services are more complex. Notably, regulators now must decide how to control the use of monopoly revenues to cross-subsidize competitive services and must decide which of the dominant carrier's services may be exempted from rate regulation altogether.

The cross-subsidy problem is particularly vexing. When a carrier's reserved services are subject to rate-of-return regulation, all costs that the carrier reports to the government as incurred to provide that service are legally recoverable through the rates charged for that service [26]. If the carrier also offers services (value-added, long-distance, or other offerings) for which it faces competition, it will be tempted to allocate some of the cost of those services to the reserved side of the ledger. Since these misallocated costs will then be recoverable through monopoly rates, they need not be recovered through the rates charged for competitive services, and those rates can be set below cost. The result is unfairness to the monopoly ratepayers and unfairness to competitors.

Regulators have fashioned different solutions to this problem. One approach is to prohibit the dominant carrier from offering competitive services;

another is to require the carrier to offer competitive services only through separate subsidiaries with their own personnel, facilities, and books of account; and another solution is to permit the dominant carrier to offer competitive services through the same organization that offers monopoly service, but require the carrier to maintain a prescribed system of cost accounting to separate the costs incurred to provide reserved services from the costs incurred to provide other services.

As the FCC has concluded in regulating the local exchange carriers' treatment of enhanced, or value-added, services, the third of these approaches permits the carriers to offer competitive services most efficiently. The first method artificially limits the number of competitors that may offer services, and the second prevents the local carriers from taking advantage of the economies of scope and scale that can be achieved through integrated operations. If the third approach has a disadvantage, it is the administrative complexity of administering a system of nonstructural safeguards. This is a game that can be played only by governments with well-staffed, well-financed regulatory agencies.

The second ratemaking problem posed by carriers offering a mix of reserved and competitive services is the problem of rate deregulation. When carriers face competition in the provision of a service, they are at a crippling disadvantage if rate regulation prevents rapid changes in price or requires prices to be set at levels that subsidize other services. The most responsible regulatory approach to this problem is to evaluate the extent of competition on a market-by-market basis and deregulate rates only in those markets in which the dominant carrier truly requires regulatory relief and will not reap any benefits from monopolistic behavior.

9.3.3 Regulating Access

As we noted earlier, the wireline, local exchange telephone network remains under the control of the dominant carriers. Until effective, large-scale bypass of these networks is feasible, therefore, all providers of competing telecommunications services will rely on the dominant carriers for the "last-mile" link between their facilities and their customers' telephones.

When the carrier controlling the local network is also a competitor of the firms that require access to that network, the potential for abuse is obvious. The local carrier may provide its competitors with technically inferior interconnections and may charge rates for interconnection that impose crippling costs not incurred by the local carrier.

Theoretically, the solution to this problem is clear enough. In order to ensure fair competition, local carriers must provide their competitors with access to the local network that is technically comparable to the access the carrier provides for its own competitive services, and they must charge reasonable, cost-based rates for those access arrangements.

While the principle of equal access is simple enough to state, achieving equal access is a complex regulatory problem. Even governments with substantial resources to devote to the task are ill-suited to specify and enforce detailed interconnection standards. The more common approach, therefore, is for the government to require dominant carriers to negotiate interconnection arrangements with their competitors in good faith. The regulatory agency will then be available to intervene if negotiations break down or commitments are not honored [27].

9.3.4 Quality-of-Service and Consumer Protection Regulation

One of the primary purposes of regulation is to protect consumers from the mistreatment monopolists can inflict—notably, price discrimination, rates that earn supracompetitive profits, and shoddy service. Where PTTs have a monopoly over all telecommunications services, all of their ratepayers are at risk for this sort of treatment. When a privatized or commercialized PTT has a monopoly over a defined class of reserved services, the customers for those services are similarly vulnerable.

The temptation to skimp on investment for basic customer service may be particularly strong under a system of incentive regulation. Since carriers make more money under price caps by reducing their costs, and since customers for reserved services may have no alternative sources for those services, it makes perfect sense to permit service to those customers to deteriorate while maintaining higher standards for those customers who have alternatives.

We have already discussed some techniques of rate regulation that protect consumers from pricing abuse. In dealing with the separate problem of shoddy service—or abandonment of service to less profitable customers—regulators have at least two alternatives. One approach is to impose a detailed set of service quality standards, either by regulation or as part of the concession through which the carrier is authorized to do business, and then to exercise continuing oversight of carrier compliance with those standards. Even when much of the oversight is exercised through carrier self-reporting, this approach is the most expensive in its use of administrative resources. Another approach is to rely on customer complaints to identify substandard carrier performance. When the regulator takes this approach, it must devote sufficient resources to the complaint process to make it meaningful. Whichever method is used, the carrier must have the authority to impose fines, withdraw service authority or otherwise punish the carrier effectively for failure to offer adequate customer service.

9.3.5 Agents and Processes of Regulation

The regulatory problems we have outlined in this section can only be solved by governments with the means to collect information, promulgate prudent rules, and resolve complaints and disagreements. The resources available for these tasks, of course, vary significantly from one country to another, as do the political structures and traditions through which the means of regulation must be devised.

In setting up a regulatory system for telecommunications, governments must accommodate a number of conflicting concerns. The regulatory body should be free of undue political interference, but it should also be responsive to overall policy direction from the legislature, cabinet, or other politically accountable sources. The regulator should also work closely with industry, but should protect the competitive process and avoid "capture" by the interests it is charged to regulate.

Some governments have tried to achieve these goals within existing governmental structures. For example, when Argentina privatized ENTel, it intended to rely on the presence of government officials on the company's board, rather than a regulatory agency, to ensure that the company would act in the public interest. Similarly, New Zealand elected to rely on its Commerce Commission, acting under its authority to enforce the general competition laws, to regulate its privatized telephone industry. Neither of these decisions appears durable: Argentina more recently decided to establish a specialized commission, and commentators in New Zealand are criticizing both the competition law and the Commerce Commission as inadequate to deal with the problems of the telecommunications industry.

Other governments have adopted specific regulatory schemes for telecommunications, but administer those schemes through politically accountable ministries rather than independent regulatory bodies. Countries taking this approach either have no specialized agencies or limit those agencies to an advisory role and concentrate licensing and other important decisions in a ministry. Examples of countries using these ministerial approaches include France, Germany, and Japan.

Other countries have adopted creative structures for sharing power between the agency and the political authorities, or between one responsible agency and another. In the United Kingdom, for example, regulatory authority is divided among a government minister, a regulatory agency (OFTEL), and a Monopolies and Mergers Commission. In Australia, the memberships of the regulatory agency (AUSTEL) and the agency empowered to enforce competition policies overlap, so that each agency will be mindful of the concerns of the other. These arrangements, among other concerns, seek to prevent capture of the regulatory agency by the industry it is supposed to regulate.

Whatever the institutional arrangements adopted, the elements necessary for effective regulation are not hard to identify. The regulator must be adequately funded and staffed; it must have credible backing from the political authorities, who must throw their weight behind controversial decisions to ensure their success; and it must have a clear, but not overly confining, legislative mandate that gives it both the authority and the flexibility to respond to conditions in a rapidly changing industry [28].

In concluding this discussion of regulation, we should point out that in practice the gap between the real and the ideal is very wide. Even well-funded, expertly staffed regulatory bodies perform unevenly and may fail to meet all of the demands placed upon them [29]. Regulators with fewer resources—especially in the developing world—may have to forego many of the tasks we have identified in favor of more immediate concerns, such as the need to attract capital and upgrade antiquated networks.

9.4 CONCLUSION

The following chapters are a partial survey, by region, of the regulatory systems in place in many of the nations of the world. The discussion contained in these chapters is not exhaustive, nor is it likely that all of it will be current by the time this book reaches the reader. Anyone needing up-to-date information on a particular country, therefore, should contact the telecommunications administration of that country or organizations such as the ITU, the OECD, and the Asia Pacific Economic Cooperation Group [30].

Notes

[1] This elastic concept means different things to different people. For an economist, an industry is a natural monopoly when its cost structure is such that a larger firm is always more efficient (i.e., has lower average costs) than a smaller firm, so that a firm large enough to supply the entire demand for the service is most efficient of all. For a regulator, it may simply mean that competition will threaten universal service or some other policy goal the regulator believes to be important. See, for example, C. Kennedy, *An Introduction to U.S. Telecommunications Law*, Norwood, MA: Artech House, 1994, p. xvi.

[2] For a description of this notion of *political telematique*, see E. Noam, "International Telecommunications in Transition," in *Changing the Rules: Technological Change, International Competition, and Regulation in Communications*, W. Crandall and K. Flamm, eds., Brookings Institution, Washington, DC, 1989, pp. 259–260.

[3] Developing countries, in particular, found that primitive telephone systems with long waits for installation of access lines retarded economic development and discouraged foreign investment.

[4] This public attitude—largely a legacy of the 1960s and the Watergate scandal—is little noted in the literature on the divestiture of AT&T, but was keenly felt by the lawyers and lobbyists who tried to make the case for the old monopoly with the public.

[5] In some cases, the PTT's telecommunications operations have been divided between two or more new entities, with the intention that they will serve distinct service markets (local service, long-distance service, international service) or geographic areas.

[6] One of the most important steps in preparing for privatization may be the accommodation of the concerns of the labor force. Another useful step, before privatization is completed, is the "rebalancing" of the PTT's tariffed rates to eliminate subsidies that will make the company vulnerable to selective competition after it is privatized.

[7] Incentive regulation is described in Section 9.3.

[8] These alternatives, along with real-world examples of their implementation, are discussed at length in B. Wellenius and P. Stern, *Implementing Reforms in the Telecommunications Sector*, World Bank, Washington, DC, 1994.

[9] This result was not accomplished by calculating the cost of each service, then deliberately setting rates for those services "above" and "below" those costs to achieve the desired rate structure. In fact, in the monopoly environment, no one knew the cost of individual services. Telephone companies simply charged as much as the traffic would bear for the subsidizing services, then set their rates for the subsidized services at a level that would recover the company's remaining costs. While no one doubted that this practice resulted in substantial internal subsidies, no one knew the precise amount of the subsidies. See, for example, Kennedy [1], pp. 23–27.

[10] Protection of the dominant carrier's more profitable services may also be intended to preserve the company's overall viability during the transition to full competition. Among other benefits, a governmental assurance that a privatized PTT will continue to enjoy a monopoly over a large class of services is a powerful incentive for investment in the company.

[11] In this context, we use *PTT* to refer to the dominant carrier, whether public or privatized.

[12] It is also contrary to the interests of long-distance companies, since a healthy local exchange carrier industry is necessary if long-distance carriers are to reach their customers.

[13] Some governments presume that services fall within the reserved category unless they are affirmatively shown to fall outside of it. Other regimes indulge the opposite presumption and require the dominant carrier to prove that a particular proposed service belongs within the reserved category.

[14] The U.S. is opening its local exchange market piecemeal, through decisions of the regulatory commissions of the individual states. See Chapter 11.

[15] In the U.S., most of the long-distance companies that compete with AT&T (although not the largest) offer their services over lines leased from the handful of large, facilities-based long-distance companies.

[16] The inanity of these distinctions, viewed from a purely technical perspective, is demonstrated by the fact that even basic telephone service, in most advanced nations, involves the electronic conversion of the analog signal transmitted over the customer's access line to a digital signal, and reconversion to an analog signal for delivery to the party receiving the signal.

[17] See Chapter 11. See also Kennedy [1], pp. 64–70.

[18] See Noam [2], p. 273.

[19] In the case of competition in the wireless telephone markets, restrictions on entry can be justified on grounds of scarcity of RF spectrum. In other markets, such as wireline long distance, the goal is to limit supply and control downward pressure on prices in order to preserve local service subsidies.

[20] As we discuss more fully in Chapter 11, local service competition in the U.S. is within the jurisdiction of state regulators.

[21] One way to accomplish this is to build the subsidies into the access charges the dominant carrier is allowed to impose on the new competitors.

[22] In the academic world, this is known as the Averch-Johnson-Willisz (A-J-W) effect. See H. Averch and L. Johnson, "Behavior of the Firm Under Regulatory Constraint," *American Economic Review*, Vol. 52, 1962, p. 1052.

[23] The basket in which residential service falls, for example, includes the separate rate elements for rental of the customer's access line, local calling rates, and long-distance calling rates.

[24] The formula is expressed as RPI - X, where RPI is the retail price index and X is the anticipated efficiency gain. So, for example, if the retail price index will rise 1% and average industrywide costs are expected to decline 5%, at the end of the review period the price cap will drop by 4%.

[25] When a company newly subject to price caps has overinvested heavily during a previous rate of a return regime, it may achieve dramatic—if temporary—profits by slashing employment rolls and otherwise becoming efficient. When this one-time program of downsizing is over, however, the downward pressure on rates built into the price cap formula may begin to pinch.

[26] The problem may be equally bad when the rates for reserved services are not regulated or are regulated ineffectively.

[27] Some regulatory authorities—notably, the EU and the FCC in the U.S.—have developed interconnection guidelines that go beyond mere statements of principles. The FCC's rules implement the rubrics of equal access, open network architecture, and comparably efficient interconnection, while the EU has elaborated ONP, which includes nondiscriminatory access, unbundling of local exchange access features, and a prohibition against cross-subsidization.

[28] Flexibility is especially important. A great source of frustration—and litigation—for the FCC has been the failure of Congress to update the Communications Act of 1934, which gives the FCC little or no discretion in removing tariffing obligations from carriers that lack market power.

[29] During the lawsuit brought by the U.S. against AT&T, witnesses for the FCC—a strong agency by any measure—protested that the Commission had been unable to supervise AT&T's rates sufficiently to ascertain its cost of service or prevent anticompetitive conduct.

[30] The following chapters emphasize domestic regulation of wireline services and pay relatively less attention to wireless and satellite services. For an excellent detailed discussion of approaches to the development and regulation of wireless and satellite technologies, see R. Schwartz, *Wireless Communications in Developing Countries: Cellular Systems and Satellites*, Norwood, MA: Artech House, 1996 (in press).

10 The Far East

Asia is the most populous, and possibly the most varied, of the Earth's continents. Its telecommunications systems range from some of the world's most advanced to some of the most primitive, and run the gamut from monopoly PTT systems to open, competitive structures subject to widely varying levels of regulatory control. This chapter considers the regulatory history and presents approaches of some of Asia's principal economies.

10.1 AUSTRALIA

Australia has followed an exceptionally deliberate course in privatizing, restructuring, and regulating its telecommunications industry. Since the Australian experience is a textbook case of many of the themes discussed in our introductory chapter, it is worth recounting in detail.

The story begins in the usual place: with all telecommunications services provided by government monopolies. In Australia's case, before 1975 all domestic telecommunications were provided by the Postmaster-General's Department, and all international services were furnished by the Overseas Telecommunications Commission (OTC).

The Australian government's first modification of the system was taken with the single goal of universal service in mind. The government concluded that in order to extend service throughout the vast Australian Outback, it needed a telecommunications administration that combined greater efficiency with the ability to maintain significant internal rate subsidies—in other words, a commercialized monopoly. Accordingly, in 1975 the government separated the PTT's postal functions from its telecommunications functions and created Telecom Australia (Telecom)—a public agency with its own capitalization, substantial administrative autonomy, and a monopoly of all domestic telephone service.

Telecom, aided by the efforts of the AUSSAT satellite system [1], achieved its goal of universal service during the decade following commercialization. By this time, the government was ready to shift its policy emphasis from basic service penetration to modernization and improved efficiency.

This second phase of reform began with a government review of telecommunications policy and a decision to move incrementally toward a privatized, competitive industry. In order to ensure effective regulation of this new industry structure, the Australian Telecommunications Authority (AUSTEL) was established in 1989 as an independent regulatory body with a broad mandate for technical and economic regulation of the industry.

Now the government was ready to begin more serious structural reform, which it pursued through the classic process of privatization accompanied by regulation of entry, pricing, and access, along with an unusually explicit set of provisions designed to ensure universal service.

The government decided to pursue liberalization incrementally, beginning with the establishment of a transitional telecommunications duopoly. The two public monopolies, Telecom and OTC, were merged in 1991 to form a new entity called Telstra, which enjoyed a temporary monopoly over all domestic and international noncellular service. The government then created a second carrier by privatizing AUSSAT through a negotiated sale to a consortium consisting of BellSouth, Cable & Wireless, and an Australian syndicate called Optus Pty Ltd., which held a 51% interest. As an inducement to investors, the government guaranteed the Optus consortium that it would face no additional facilities-based competition—that is, no competitors other than the Telstra system—in noncellular services until mid-1997.

This duopoly system is scheduled to end on June 30, 1997, with full competition in all markets permitted after that date. In the meantime, full competition is permitted in value-added services, customer premises equipment, mobile telephone service, and resale of domestic and international facilities.

Along with this liberal scheme of entry regulation, Australia has implemented a substantial system of pricing and access regulation. On the pricing side, Australia adopted an incentive system, based on price caps, similar to the approaches taken in the United Kingdom and several U.S. jurisdictions. On the access side, Telstra must provide (and before its creation was required to certify its readiness to provide) equal access for Optus to the Telstra network. Specifically, Telstra must offer 1+ dialing parity and efficient access arrangements at rates that recover only the direct, incremental cost of those arrangements.

Finally, Australia has put in place an explicit set of universal service requirements. Specifically, the government has recognized three categories of so-called community service obligations (CSO): universal service (access to basic residential service and pay telephones), emergency services, and discounted service for the disabled and charitable organizations. The obligation to provide these services at affordable rates is assigned to Telstra, but the cost of supporting the provision of CSOs is shared between Telstra and Optus.

While the Australian experience is exemplary in many ways, the present system has its anomalies. Chief among them is the continued government ownership of the dominant carrier. It remains to be seen whether a nonprivatized entity will achieve the efficiencies required in the modern telecommunications environment and whether the government will at all times act impartially in regulating a market in which it is, itself, the principal player.

10.2 CHINA

In 1949, shortly after the Communist takeover and the founding of the People's Republic, telecommunications service in China was centralized in the Ministry of Posts and Telecommunications (MPT) [2]. The result was a rigid, hierarchical regime for the regulation of all telecommunications products and services offered in the People's Republic and between the People's Republic and foreign nations.

The central government of China has not retreated, in principle, from the notion that the MPT is the monopoly regulator and provider of telecommunications services for all of China. Since the general policy of economic liberalization began in 1979, however, significant reforms have substantially diluted the effective, if not the official, control of the MPT over investment and competition in the industry. This process has been aided by two long-standing features of China's telecommunications infrastructure: the existence within the MPT of local and regional administrations that have come to operate with increasing autonomy, and the extensive development, by ministries other than the MPT, of separate telecommunications networks serving the electrical, transport, and other sectors of the state-run economy.

The power of these entities to chart their own destinies increased dramatically in the early 1980s when the central government, which had been the exclusive source of funding for telecommunications investment, determined that new sources of financing would have to be tapped if services and facilities were to meet growing demand. Accordingly, the government authorized the local authorities, as well as national ministries operating independent telecommunications systems, to obtain financing from local governments and private sources, and to enter into joint ventures with foreign entities [3]. The result has been an explosion of new ventures—many involving non-Chinese partners—of which a few examples must suffice.

In December of 1993, China's State Council approved the creation of a new, independent telephone network operator based on the systems of the electronics, electrical power, and railway ministries. The new entity, called China Unicom, is a state-owned corporation chartered to obtain its own financing and bear its own profits and losses. While its principal aims are to offer mobile and value-added services, it plans to carry substantial long-distance telephone traffic as well. BellSouth, GTE, and Sprint have signed agreements to assist in the deployment of this "second network" [4].

Northern Telecom is participating in at least five ventures undertaken by local telecommunications administrations. One contract, with the administration of Liaoning Province in northeast China, involves high-speed intercity digital transmission facilities; another, with Hebei Province, is for the provision of Northern Telecom digital switches. Other Northern Telecom contracts are for a semiconductor chip manufacturing venture in Shanghai, switching, transmission, cellular, and fiber-optic installations in Guangdong Province, and a switch manufacturing plant (also in Guangdong). Northern Telecom's partners in these ventures include the Guangdong Provincial PTT, the Hebei PTT, the Henan PTT, and a number of corporations formed by the local administrations [5].

As of the fall of 1994, AT&T had formed nine joint ventures with local administrations to develop advanced telecommunications facilities in China. Notably, the Posts and Telecommunications Bureau of Guangdong Province formed a limited partnership with AT&T to develop advanced communications in Southern China, including digital transmission systems and switches [6].

So far, all of this ferment has resulted in a system in which national and local bureaucracies, all under the official control of the MPT, team up with private (often foreign) partners to compete in the markets for manufacturing, long-distance, and wireless services. (Local wireline service is still a monopoly.) The entire process remains a governmental enterprise: the local authorities and ministries are the majority owners of each of the new ventures, and China retains—for the moment at least—an outright ban on foreign ownership of telecommunications service companies. At the same time, the MPT retains its prerogative to control pricing of basic services, standards, and the overall development of the telecommunications infrastructure, including the implementation of the telecommunications phase of the central government's ambitious ten-year economic and social development plan (1991 to 2000).

The Chinese system is, to say the least, a peculiar hybrid of wide-open competition within a governmental structure designed at a time of totalitarian control over the social and economic life of the nation. It remains to be seen how durable and how effective in meeting the telecommunications needs of the largest nation on Earth this system will be.

10.3 HONG KONG

Hong Kong, like Singapore, is a city-state with an advanced economy and none of the universal service problems of nations with large rural hinterlands. Accordingly, its telephone penetration rate is among the highest in the region.

At this writing, facilities-based, noncellular telecommunications services in Hong Kong are provided by private monopoly companies. Hong Kong Telecommunications Ltd. owns both the domestic service provider, Hong Kong Telephone Company, and the international provider, Hong Kong Telecom International [7]. The government intends, however, to permit competition in internal telephone service beginning in 1996, after the expiration of Hong Kong Telephone Company's exclusive franchise [8]. (As part of its preparation for the competitive era, Hong Kong Telephone has elected to change from rate-of-return regulation to price cap regulation.)

The government has created a regulatory agency called the Office of the Telecommunications Authority (OFTA). The new authority's mandate includes intervention in disputes that arise between Hong Kong Telephone and new entrants concerning the terms of access to Hong Kong Telephone's network, and the administration of the dominant provider's ongoing universal service obligations.

Markets for mobile and value-added services in Hong Kong are already competitive, and some competition has also emerged in the provision of international value-added networks (IVAN). Once the new local franchisees begin operation in 1996, therefore, international voice and basic data services will be the last holdout of monopoly services in Hong Kong.

Looming over all political and economic arrangements in Hong Kong, of course, is the scheduled takeover by the People's Republic of China in 1997. Fortunately, China is increasingly liberalizing its own economy (although its political system is still quite oppressive) and has taken a financial stake in Hong Kong Telecom. If China acts in its economic interest, therefore, it will benefit from Hong Kong's continuing efforts, in competition with Tokyo and Singapore, to become the dominant trading and financial center in Asia.

10.4 INDIA

India, with a population of nearly one billion people, is one of the largest nations on Earth. Its domestic telecommunications service is relatively primitive, with a telephone penetration rate of less than 1% and long waiting times for service.

Compared with Japan and China, India has made scant progress in opening its telecommunications markets to competition and securing capital for modernization. Except in the New Delhi and Bombay areas, local and domestic long-distance services are still provided exclusively by the Department of Telecommunications, a nonprivatized government agency that acts both as regulator and service provider. In Bombay and New Delhi, local service is provided by Mahanagar Telephone Nigam Limited (MTNL)—a carrier in which the government holds a majority stake. International long-distance is provided by yet another monopoly, Videsh Sanchar Nigam Limited (VSNL), which is also mostly government-owned.

The Indian government has taken some steps toward increasing private investment in the monopoly service providers and has sponsored serious studies of the alternatives for further reform. One proposal, made by a government committee in 1991, was to decentralize the Department of Telecommunications (DOT) bureaucracy by splitting it into five regional operating companies and a separate domestic long-distance company. This proposal, reminiscent of the Chinese approach, also envisioned separation of the DOT's regulatory functions from its operating functions. This set of recommendations, like subsequent proposals for private cellular networks and a modern user-supported "overlay" network for business customers, foundered in the face of controversy, litigation, and political opposition. In fact, in spite of intense study and discussion of the need for reform of the telecommunications service sector, the most sustained effort to date has been the government's attempt to encourage a domestic equipment manufacturing industry. As part of this program, considerable government resources have been invested in the production of modern switching systems for use in India's rural exchanges—many of which still switch calls manually.

The many difficulties cited here notwithstanding, signs of progress in the Indian telecommunications sector can be found. For one thing, in India, as in China, other infrastructure agencies of the government—including the National Railways and the Oil and Natural Gas Commission—are developing their own, parallel telecommunications networks for their internal use. These networks may form the basis for competitive provision of facilities-based services. And the government also has approved a joint venture, involving U S West and an Indian partner, to compete with DOT for local service in the city of Tirupur. Most recently, the Indian government has announced its intention to privatize its telecommunications operations and award licenses to private providers of basic telephone service. In early 1995, a new three-member Telecommunications Regulatory Authority was created to screen license applications and review the performance of licensees.

10.5 INDONESIA

Indonesia is a collection of 13,600 islands (6,000 of which are inhabited) extending along the equator for a distance of about 3,000 miles. Telephone penetration is less than 1%, and 75% of Indonesia's villages have no telephone service of any kind.

Reform of Indonesia's telecommunications sector has proceeded only as far as the corporatization—but not the privatization—of the PTT. All domestic noncellular service is provided by PT Telekomunikasi Indonesia (PT Telkom), and most international services are provided by PT Indosat. Both carriers are 100% government-owned.

Private investment and competition are permitted only in mobile telephone, satellite, and international services. Notably, PT Satelindo, a corporation in which PT Telkom holds a 30% interest, PT Indosat holds a 10% interest, and private investors hold a 60% interest, operates the Palapa satellite system and is licensed to provide cellular and international service.

While the government's chief priority for the sector is to increase telephone penetration, until recently it resisted the notion of permitting private investors (and foreign investors in particular) to acquire a controlling interest in any provider of basic service [9]. More recently, however, the government announced its intention to privatize PT Telkom in the fourth quarter of 1995.

10.6 JAPAN

Japan is one of the world's largest and most advanced economies and has correspondingly sophisticated systems of telecommunications and regulation. This section describes the history of telecommunications regulation in Japan, and then describes how Japan handles the common regulatory problems of entry, pricing, foreign ownership, and control of market power.

10.6.1 The History of Regulation in Japan

Japan's telephone and telegraph system was established on the European model, as a government monopoly within the same agency that ran the postal system [10]. In the early 1950s, this system was replaced by the creation of two monopoly corporations: the Nippon Telephone and Telegraph Corporation (NTT), which was government-owned and would provide all domestic services, and the Kokusai Denshin Denwa Company (KDD), a private monopoly carrier of international services [11]. From 1953 until the privatization initiative of 1985, these companies were insulated from competition by law (even private networks and inhouse telephone systems were forbidden) [12], and the introduction of new products and services was often considerably delayed [13].

In 1980, the Japanese government began to study the impact of regulated monopoly on the development of the nation's telecommunications infrastructure, and concluded that the prohibition on new entry was impeding needed innovation. The government determined that introducing competition and privatizing NTT would spur technological progress and (as the Ministry of Finance earnestly pointed out) bring much-needed funds to the public treasury [14]. The reform process began when the National Diet passed the Telecommunications Business Act and the NTT Corporation Act. The former established a new regulatory structure for the Japanese telecommunications industry, and the latter converted NTT into a shareholder-owned company with the Japanese government holding one-third of the shares. As a result of the 1985 initiatives, both international telecommunications and internal long-distance service are now competitive, as are cellular telephone service and telecommunications equipment markets. NTT continues to operate as a monopoly carrier for local telephone service, however, under a regime of close regulatory scrutiny by the MPT [15].

10.6.2 Regulation in Japan Today

The 1985 reforms created two classes of Japanese telecommunications companies—Type 1, or facilities-based, carriers, and Type 2, or resale, carriers. The Telecommunications Business Act divided Type 2 carriers into two additional categories: Special Type II carriers, which offer service to the public generally or between Japan and foreign destinations; and General Type II carriers, which offer services to defined user groups rather than to the public at large [16]. Both types of carriers may provide any kind of telecommunications service they choose (except for local exchange service, which is provided exclusively by NTT).

Entry, pricing, and service regulation vary substantially from one category of service provider to another.

Type I carriers, as we might expect, are subject to the most stringent regulation. They are the only carriers with universal service obligations; that is, they must provide service to all who request it at an acceptable level of quality. For this reason, the MPT scrutinizes the qualifications of businesses seeking to do business as Type I carriers and may reject those applicants it finds to be unqualified. All of the Type I carriers' terms of service must be published in tariffs filed with the MPT, they must have the MPT's permission before discontinuing any service, and they must set out their plans for expansion of facilities in plans filed annually with the MPT [17].

Type II carriers are subject to a less formal scheme of regulation. Entry regulation is by registration with the MPT in the case of Special Type II carriers, and simple notification to the MPT in the case of General Type II carriers. (While the MPT has the authority to reject registrations of Special Type II carriers that appear unqualified under criteria set out in the Act, this authority seems to have had little effect on the growth of this segment of the industry [18].) Special Type II carriers offer their services under tariff, but MPT approval of those tariffs is not required. General Type II carriers need not file tariffs at all. Type II businesses, unlike Type I carriers, are not subject to foreign ownership restrictions.

While liberalization of entry, price, and foreign investment regulation appear to have gone smoothly since 1985, two issues continue to present more fundamental problems: the continuing monopoly power of NTT in local telephone markets and the opening of Japanese markets to effective foreign competition. We consider the first of these issues in the following section.

Coping With the Market Power of NTT

In the years since 1985, the MPT has made a positive effort to assist companies seeking to compete with NTT. Notably, the ministry has held NTT's long-distance rates some 10% to 20% higher than those of its competitors, has required NTT to provide technical assistance to its competitors, and has given expedited consideration to the requests of new Type 1 carriers to build facilities [19].

Despite these efforts, NTT continues to enjoy the structural advantage that AT&T enjoyed in the United States before divestiture: it offers long-distance service in competition with other carriers, yet controls the access of other long-distance companies to the local exchange networks. In the U.S., this problem was addressed by divesting AT&T's local operations from its long-distance network and requiring the divested local companies to offer equal access to all long-distance carriers [20].

NTT's competitors have complained of discriminatory conduct of the kind charged against the predivestiture AT&T. Notably, customers served by NTT's older, nonelectronic switches cannot subscribe to competing carriers at all, and those who do subscribe to competitors must dial more digits than NTT's customers to reach their carriers. Competitors also complained that NTT could use its relationship with the competitors' customers as their local carrier, and its role in changing customers over to competing carriers, to acquire and abuse marketing information to the detriment of the competitors.

The National Diet has the power, of course, to amend the Telecommunications Business Law and order divestiture of NTT's local operations [21]. In 1990, a report commissioned by the MPT recommended that NTT divest itself of its long-distance operations, with each NTT shareholder to receive one share of stock in each of the new companies, and that NTT spin off its cellular and paging operations into a separate company as well. The report also recommended that the divestiture of mobile services take place in the 1991 fiscal year, and that the divestiture of long-distance services take place in 1995. The delay in long-distance divestiture was intended to permit NTT to complete its transition to digital transmission technology [22].

The MPT did not adopt the committee's divestiture proposals—largely because of opposition from business, labor, and NTT shareholder interests [23]—and merely ordered NTT's mobile business to be transferred to a separate subsidiary, and the establishment of a separate division for long-distance service. The MPT then decided to defer the question of divestiture until 1995, when the MPT would assess the telecommunications marketplace and determine whether divestiture was necessary. In 1995, on schedule, the MPT reopened its inquiry into the possible divestiture of NTT, and in the fall of 1995, the MPT's advisory group recommended splitting NTT into five regional companies with authority to compete in each other's territory. The Ministry is expected to make a final decision on these recommendations in February 1996.

In the five years between 1985 and 1990, NTT itself has taken steps to improve its competitive image by voluntarily divesting its cellular operations and improving its interconnection and customer information policies to accommodate its competitors' complaints [24]. Notably, in March of 1995, NTT announced a plan to open its local network to expanded interconnection with other carriers. The announced innovations include:

1. Streamlined processing of interconnection requests, with standardized documents and response times;
2. Special switches to interconnect NTT's signaling network with those of other carriers;
3. Appropriate access charges, including charges to other carriers for network upgrades needed in order to accommodate their interconnection requests [25].

Whether NTT's voluntary efforts will persuade the MPT and the National Diet to forego divestiture of the company, of course, remains to be seen.

The MPT, like the FCC in the United States, is also confronting the question of subsidized rate structures. NTT's long-distance competitors, like MCI in the early days of long-distance competition in the United States, have concentrated on service between the more profitable sets of city-pairs and has undercut NTT's long-distance rates substantially. NTT has indicated a desire to redress this imbalance by increasing its local rates, which it contends are subsidized by its long-distance rates. The MPT has not yet determined whether it will permit the subsidized rate structure of NTT to be maintained.

10.7 KOREA

Korea, like Australia, has followed a multistage program of telecommunications reform that began with a government-funded effort to increase telephone penetration [26]. This first phase of the program achieved a fivefold increase in access lines in only 10 years [27].

The second phase began with the commercialization of Korea's PTT through the creation of Korea Telecom, a 100% government-owned corporation, and the enactment that same year of a new telecommunications law. (Plans to privatize Korea Telecom were announced in 1995.) The 1991 law established three categories of service providers: General Service Providers for wireline telephony, Special Service Providers for wireless services, and Value-Added Service Providers. The law also established a framework for new entry into the telecommunications sector.

At this writing, Korea has a partial duopoly in the General Services market, with a more liberal entry policy in wireless and value-added services. The duopoly approach was introduced in 1991 when Data Communications Corporation of Korea (DACOM), a private corporation formed in 1982 as a monopoly provider of data communication services, was permitted to provide international voice services in competition with Korea Telecom [28]. DACOM hopes to expand the duopoly by securing authorization to offer domestic long-distance voice services.

In the meantime, except for restrictions on the permissible percentage of foreign ownership, entry regulation for wireless and value-added services has been substantially liberalized. And in order to ensure a competitive mobile market, Korea Telecom and DACOM may not offer wireless telephone services.

10.8 MALAYSIA

Malaysia is composed of two territories—East Malaysia and West Malaysia—separated by 400 miles of open sea. East Malaysia lies along the north coast of Borneo, and West Malaysia, which contains most of the country's population, occupies a peninsula between Thailand to the north and Singapore to the south.

Although telephone penetration in Malaysia is low by the standards of Singapore, Hong Kong, or Japan, the economy is developing rapidly and the government is committed to achieving a penetration rate of 40% by the year 2005. Consistent with this goal, the government has corporatized and partially privatized its PTT and is licensing competitive service providers in all segments of the market.

Corporatization began in 1987 when the government created Syarikat Telekom Malaysia Berhad (Telekom Malaysia) as a government-owned corporation and assigned to that company the telecommunications service functions formerly performed by the Ministry of Energy, Posts, and Telecommunications. In 1990, the government began the privatization of the new corporation by offering 23% of its shares for sale [29]. (In 1995, Telekom Malaysia announced its intention to split itself into three divisions, with one providing basic service and the other two providing new and enhanced services.)

More recently, the Malaysian government has issued a number of licenses for cellular, satellite, and facilities-based voice telephone services. Some of the licensees will offer the dominant carrier robust competition. Technologies Resource Industries (TRI), for example, bought Telekom's cellular operation and plans to become an international services provider. Binariang Sdn Bhd, a company with substantial nontelecommunications holdings, has a local telephone service license and is authorized to launch and operate the MEASAT satellite. And a number of licensees are operating, or have announced their intention to operate, competing cellular, international, and long-distance telephone services.

The Malaysian telecommunications sector is regulated by a specialized bureau within the Ministry of Energy, Posts, and Telecommunications, but no detailed set of rules is in place for regulation of the new competitive environment. This appears to be a matter of choice: the government apparently intends to make rules for the sector as experience suggests the need for them, rather than on an a priori basis.

10.9 NEW ZEALAND

Reform of New Zealand's telecommunications sector began in the 1980s and has resulted in an industry that is open, competitive, and—at this writing—thinly regulated. Prior to the 1980s, New Zealand's government owned and operated a wide range of commercial activities, ranging from the post office and the telephone system to banking and coal mining. The telecommunications network, which was run by the post office, enjoyed a monopoly of all domestic and international services. Unfortunately, by the time of a 1986 government report, neither those services nor any of the other state-run enterprises was showing a profit.

The first stage of reform—the corporatization of the PTT—began in 1986. The PTT's telecommunications functions were assumed by a government-owned corporation called Telecom Corporation of New Zealand (TCNZ), which subsequently was privatized through a negotiated sale to a consortium made up of Bell Atlantic, Ameritech, and two New Zealand corporations.

The privatization of TCNZ was unusually complex. While Bell Atlantic and Ameritech initially bought a majority interest in the company, they were required to make a public offering to the New Zealand public and reduce their interest to 49.9% or less within three years of privatization [30]. They were also required to guarantee the government's retention of a so-called "Kiwi share," which gives the government voting rights intended to ensure TCNZ's adherence to foreign ownership restrictions and the service obligations contained in its articles.

Competition is permitted in all telecommunications services in New Zealand, but no separate regulatory agency has been established to supervise the industry. Instead, New Zealand relies on its Commerce Commission to regulate the industry through the general competition laws [31] and the provisions of the Telecommunications Act of 1987. The relevant Telecommunications Act provisions require TCNZ to disclose its rates and other terms of service and to maintain separate books for its regional operating companies [32].

So far, at least, New Zealand has not introduced detailed regulations for pricing, interconnection, universal service, and service quality. The government has, however, exercised its "Kiwi share" rights to impose service obligations on TCNZ, including geographic rate averaging and limitations on residential rate increases. Other traditional regulatory concerns, such as the control of discriminatory access and similar abuses of TCNZ's dominant market position, must be addressed through application of the competition laws. The adequacy of this approach, of course, remains to be demonstrated.

10.10 PAKISTAN

Pakistan is a largely rural nation of over 100,000,000 people with a telephone penetration rate of slightly less than 2%. Its efforts at sectoral reform have yielded a corporatized PTT and competitive markets for paging, cellular, and pay telephone services, but do not yet include privatization of the PTT or the opening of basic markets to new entry.

The principal service provider in Pakistan is the Pakistan Telecommunications Corporation (PTC), which was formed in 1991 when the Pakistan Telephone and Telegraph Department became a 100% government-owned company. Unlike most countries that have corporatized their PTTs, however, Pakistan did not relieve the new corporation of the regulatory obligations of the former PTT. Accordingly, PTC both provides services and regulates the industry.

The Pakistani government has recognized the need to privatize PTC, open basic markets to competition and establish an independent regulatory body. While the government's efforts in this direction have been frustrated by political obstacles, Prime Minister Bhutto announced in late 1994 that the government had decided to begin the process by selling 26% of PTC to a "strategic partner" already selected by the government.

10.11 THE PHILIPPINES

The telecommunications sector in the Philippines is unlike any other system we have discussed. Most telephone service is offered by a single provider—a private company called Philippine Long-Distance Telephone Company (PLDT)—with a market share more than sufficient to confer monopoly power. At the same time, many communities are served by independent telephone companies, some of which are not even interconnected with the PLDT network [33]. This fragmented industry has achieved low penetration rates and generated long waiting times for service, even in Manila. Conditions elsewhere in the archipelago of 7,100 islands are worse.

Competition has emerged in a number of nonbasic markets, including international, cellular, and satellite service (both domestic and international). Competition in domestic, switched telephone service is not prohibited, but PLDT has frustrated the plans of would-be competitors in these markets by resisting demands for interconnection with its network.

In some ways, the situation in the Philippines resembles that of the United States before the divestiture of AT&T, with competition appearing in niche markets and the dominant, private provider hindering the attempts of new carriers to enter its core business. Unlike the United States in the 1970s and 1980s, however, the Philippines must find a way to encourage competition while at the same time improving the dominant carrier's performance and finding the capital needed to improve its service to urban customers and extend service to the vast, unserved segment of the population.

Unfortunately, the regulatory tools needed to accomplish these tasks are not immediately available. Telecommunications services are subject to the overlapping, ill-defined jurisdictions of two agencies: the National Telecommunications Commission, which handles licensing and radio spectrum allocation, and the Department of Transportation and Communications, which has a mandate broad enough to include regulation of access, pricing, and service quality. These agencies are badly in need of legislation to allocate responsibility between them and provide policy direction in carrying out their responsibilities.

10.12 SINGAPORE

Singapore, like Hong Kong, is a prosperous city-state with no rural areas to serve and a highly developed telephone system. It operates a highly successful telecommunications sector without benefit of full privatization or open markets and therefore has scant incentive to change its approach.

The dominant provider is Singapore Telecom (ST), which was corporatized in 1992. Substantial interests in ST have been sold to nongovernment investors, but the government retains (and intends to continue to retain) majority control. ST has a monopoly in domestic services, but will relinquish its monopoly over international services in 2007 and will give up its monopoly over mobile services in 1997. In the meantime, competition is permitted in IVANs so long as license applicants can show that they add substantial value and are not merely reselling ST facilities.

At the moment, ST is both the monopoly service provider and the effective regulator of the telecommunications sector in Singapore [34]. Under these conditions, and given the consistently high financial and service performance of ST, there is no strong mandate to subject ST to competition. If such pressures do arise, they are likely to come from ST's own desire to offer services in foreign markets and the reciprocal demands of ST's trading partners for access to Singapore's markets.

10.13 TAIWAN

Basic telecommunications services in Taiwan are still provided by a monopoly, noncorporatized public agency that both operates and regulates the system. The principal provider is the Directorate General of Telecommunications, which is a bureau of the Ministry of Transportation and Communications. New entry is permitted only in specific value-added services, including information storage and retrieval and e-mail. Taiwan also maintains an absolute prohibition on foreign investment in any telecommunications service provider, including providers of value-added services.

As in Singapore, pressure to change the monopoly environment is muted by the relative success of the PTT in achieving a high penetration rate. Nonetheless, legislation has been introduced that would provide for some foreign investment in the sector and for competitive provision of mobile services and a wider range of value-added services. No serious legislative proposal, however, has contemplated liberalization of the basic wireline telephone markets.

10.14 THAILAND

Thailand's telecommunications sector is also dominated by government-owned service providers. Specifically, domestic services are provided by the Telephone Organization of Thailand (TOT) and international services are provided primarily by the Communications Authority of Thailand (CAT). Other services, including telex and packet switching, are furnished directly by a bureau of the Ministry of Transport and Communications.

Thailand lacks a clear policy concerning the scope of permitted competition. While no competitive service providers have been licensed for basic telephone service, the government-owned carriers have entered into concession arrangements with private companies to develop mobile telephone and data communication services. In addition to these efforts, the government has sought to involve private companies in basic service markets through build-transfer-and-operate (BTO) arrangements, under which a contractor funds and builds a network in exchange for authority to operate it for some period of time before returning control to the government entity. One such BTO, involving a Thai corporation and NYNEX, will build three million new access lines for telephone service in and around Bangkok.

Recently (March 1995), Thailand's government announced a plan to permit 49% nongovernment ownership of TOT and CAT, with foreign ownership limited to 20%. If Thailand is to continue to attract capital—particularly foreign capital—to its telecommunications sector, it will need to make some political and structural reforms. Notably, investors will be more likely to put their money in Thailand if the country continues to exhibit political stability [35] and separates the regulation of its industry (currently in the same ministry that controls TOT and CAT) from the ownership and management of the dominant service providers.

Notes

[1] AUSSAT, a government-owned domestic satellite system, was established in 1983.

[2] See L. Sun, "Telecommunications Development in China: Problems, Policies, and Prospects," in *Global Telecommunications Policies: The Challenge of Change*, M. Jussawalla, ed., Westport, CT: Greenwood Press, 1993, p. 172.

[3] Ibid., pp. 177–178.

[4] *Telecommunications Reports*, October 17, 1994, pp. 23–24.

[5] *Telecommunications Reports*, November 14, 1994, p. 21.

[6] *Telecommunications Reports*, September 5, 1994, p. 21.

[7] The majority owner of Hong Kong Telecommunications Ltd. is Cable & Wireless plc.

[8] Three new domestic service franchises were awarded in 1993. In the meantime, the monopoly provider's exclusive international franchise is scheduled to expire in 2006.

[9] The 1989 telecommunications law permits private investment in telecommunications, but requires the government to control the sector and permits competition only in nonbasic services.

[10] In 1948, the postal service was separated from telecommunications when the Ministry of Communications was reorganized into a Ministry of Postal Services and a Ministry of Telecommunications. T. Tomita, "Telecommunications Policy in Japan," in *The Telecommunications Revolution*, H. Sapolsky et al., eds., London: Routledge, 1992, p. 114.

[11] This change was wrought by the Wire Telecommunications Law, the Public Telecommunications Law, and the Nippon Telegraph and Telephone Public Corporation Law. Ibid., p. 116.

[12] Ibid., p. 115.

[13] Telephonic computer communications and attachment of customer-provided equipment to the network were allowed some 10 years later in Japan than in the United States. Ibid., p. 116.

[14] At the time, the National Railways alone were running an annual deficit of 30 trillion yen. Tomita [10], p. 116.

[15] See H. Oniki, "Impacts of the 1985 Reform of Japan's Telecommunications Industry on NTT," in *Global Telecommunications Policies: The Challenge of Change*, M. Jussawalla, ed., Westport, CT: Greenwood Press, 1993, p. 171.

[16] Specifically, Special Type II carriers either provide international service or control 500 or more circuits. Ibid., p. 70.

[17] Type I carriers are also subject to foreign ownership restrictions; specifically, foreign investment in these carriers is limited to one-third, and no foreign investment is permitted in NTT and KDD. See N. Mutoh, "Deregulation of Japan's Telecommunications Business and the Role of Kokusai Denshin Denwa," in *International Communications Practice Handbook*, Federal Communications Bar Association, 1993.

[18] Oniki [15], p. 70.

[19] See Oniki [15], pp. 72–73. In 1989, when carriers initiated international service in competition with KDD, the MPT held a similar price umbrella over those new carriers.

[20] See discussion in Chapter 11.

[21] The 1985 Telecommunications Business Law provides for its possible revision after three years (i.e., after 1988), and the NTT Corporation Act permits a possible reorganization of NTT after five years.

[22] See *Telecommunications Reports*, March 5, 1990, pp. 1–2. An interim report had suggested three possible scenarios: one would separate NTT into one long-distance company and one independent nationwide local carrier, another would separate NTT into a long-distance carrier and several local companies, and a third option would split NTT into a number of regional entities providing both local and long-distance service. Oniki [15], p. 74.

[23] The most important shareholder was the Japanese Government, which still owned two-thirds of NTT's stock in 1990.

[24] NTT turned over its enhanced, or computer communication, services to a new Type II company in 1988.

[25] *Telecommunications Reports*, March 13, 1995, p. 41.

[26] Our discussion here concerns the Republic of Korea and excludes the bizarre garrison state lying north of the 38th parallel.

[27] Korea had 2.8 million access lines in 1980 and over 15 million access lines in 1990. See R. Bruce and J. Cunard, "Restructuring the Telecommunications Sector in Asia: An Overview of Approaches and Options," in *Implementing Reforms in the Telecommunications Sector*, B. Wellenius and P. Stern, eds., World Bank, Washington, DC, 1994.

[28] DACOM was also required to give up its monopoly of data services, and Korea Telecom was required to relinquish its partial ownership interest in DACOM.

[29] Part of the offering was made on the Malaysian stock exchange as part of a deliberate effort to develop the domestic capital market.

[30] This condition was satisfied as of September 1993.

[31] The principal competition law is the Commerce Act of 1986, which in quite general terms prohibits abuse of market power and collusive practices that restrain competition.

[32] The Act also imposes accounting and other requirements on international carriers operating in New Zealand.

[33] There are about 70 such local networks, some of which are government-owned and in the process of privatization.

[34] An independent regulator—the Telecommunications Authority of Singapore—was created in 1992.

[35] A military coup in 1991, for example, stalled the BTO arrangement in the Bangkok area and was partly responsible for a substantial restructuring of that project.

North America

11

The North American continent stretches from above the Arctic Circle to below the Tropic of Cancer, and from Greenland in the east to Alaska in the west. Its principal nations are Canada, Mexico, and the United States of America, but the continent also includes Bermuda and the French overseas department of St. Pierre and Miquelon.

Culturally, politically, and economically, North America is a diverse continent. The United States is the world's largest economy with the oldest continuous republican government of all the world's great nations. Canada, the second-largest nation in the area, is a self-governing British dependency and an advanced industrial nation, with most of its population concentrated within 100 miles of the border with the United States. Mexico, a Spanish-speaking republic with strong cultural ties to Latin America and equally strong economic ties to the United States, is less advanced industrially than Canada and the United States, but contains immense reserves of oil and natural gas and an increasingly vibrant manufacturing sector.

With the passage of NAFTA, the Canadian, Mexican, and U.S. economies are linked in one of the world's freest and wealthiest trading blocs. All three countries have also liberalized their telecommunications sectors, and the following sections describe those efforts in some detail.

11.1 CANADA

Canada is a country of great distances, low population density, and an often harsh climate. Ubiquitous communications are essential to the country's economic development and political unity, and Canada's telephone penetration rate is one of the highest in the world.

For most of its history, telecommunications service in Canada has been provided by a mixture of private and public carriers, many of which serve individual provinces. Bell Canada (a private company), for example, serves Ontario, Quebec, and part of the Northwest Territories; public companies serve Manitoba and Saskatchewan; and private carriers cover Alberta, British Columbia, and the Maritime Provinces. In order to coordinate the operations of these companies into an efficient, nationwide network, these companies formed an unincorporated association called the Trans-Canada Telephone System, now known as Stentor [1]. Like the U.S. communications sector, the system until recently was a geographic patchwork of regional monopolies.

Canada also resembles the United States in that it has a federal system in which the central government shares power with the provinces. Canada's method of sharing authority over telecommunications has been somewhat different from the U.S. system, however: In Canada, until 1989 the federal government had plenary jurisdiction over carriers in Ontario, Quebec, and British Columbia, and no jurisdiction whatever over carriers in other provinces. A 1989 court decision gave the central government authority over interprovincial communications, and the Telecommunications Act of 1993 established a new framework in which most facilities-based carriers are subject to regulation by the Canadian Radio-Television and Telecommunications Commission (CRTC).

The opening of Canada's telecommunications sector to competition has been a gradual process, culminating in the reforms of the 1993 Act. Liberalization started in 1979, when the CRTC permitted a competitive long-distance company, Unitel Communications, Inc., (then known as CNCP) to interconnect with the Bell Canada network [2]. Subsequently the CRTC authorized mobile services providers to interconnect with long-distance networks, permitted resale and sharing of private line service, and authorized resale of U.S. and overseas traffic.

The 1993 Act established a regime that essentially deregulates resale carriers and gives the CRTC flexibility in its choice of regulatory approaches to facilities-based carriers. Common carriers under federal jurisdiction that own or operate their own facilities are called in the Act "Canadian carriers," and include Telesat, Teleglobe (the international service provider), Unitel, and most of the regional carriers belonging to the Stentor group. These carriers are required to be 80% Canadian-owned.

At the time of this writing, local wireline service in Canada is still a monopoly, although the CRTC is implementing an open regulatory framework policy that will eventually open local markets to competition. Long-distance service is fully competitive, cellular services are provided on a duopoly basis, and all services requiring radio spectrum are subject to regulation under the Radiocommunication Act.

11.2 MEXICO

Mexico is a nation of nearly 100,000,000 people with a telephone penetration rate of slightly under 10%—a fairly typical penetration rate for Latin America. Forty percent of the telephones are in Mexico City, and installation of a telephone requires a wait of from 12 to 18 months. The government's goal is to cut this waiting time in half and raise the nationwide penetration rate to between 20% and 30% by the year 2000.

Reform of the Mexican telecommunications sector began in the mid-1980s, when a worldwide decline in oil prices caused an economic crisis and halted the expansion program undertaken by the government-controlled PTT, Telefonos Mexico (TelMex) [3]. Responding to the crisis, the government of President Salinas de Gortari embarked on a program to privatize TelMex and expand and modernize the telecommunications network.

The privatization program began in 1988, when TelMex was reorganized into a more efficient structure consisting of seven directorates with various functional and regional responsibilities, and with each directorate accountable for its own financial performance. Concurrently with this corporatization effort, the government announced a privatization program and assured the public that Mexican investors would maintain majority control, that ultimate regulatory oversight of the telecommunications sector would remain with the government, that the rights of TelMex's labor force would be protected, and that the privatized company would operate under a concession containing stringent quality-of-service standards.

In preparation for the sale of TelMex, the company's management developed and in 1990 the shareholders approved a new capital structure for the company. So-called AA shares, which had full voting rights, could be owned only by the Mexican government and by Mexican persons and corporations. (AA shares had previously been owned only by the government, and were of course the majority of the company's capitalization.) Another class of shares, called A shares, also had full voting rights but could be owned by anyone, Mexican or foreign. And a third class, called L shares, had limited voting rights and no ownership restrictions. The shareholders agreed that 20.4% of the privatized company's capitalization would be in AA shares, 19.6% would be in A shares and 60% would be in L shares.

The shares were sold in three stages. First, the government sold 4.4% of its AA shares to TelMex's employees. Next, the entire bloc of A shares was auctioned to bidding groups in which foreign interests could participate, but in which Mexican concerns were required to constitute the majority. The auction resulted in a sale to a consortium composed of Grupo Carso, a Mexican company, France Telecom, and Southwestern Bell. Finally, L shares were placed in public and private offerings in capital markets around the world, including the New York Stock Exchange.

The privatized TelMex acquired its authority to operate from a new "Title of Concession" that gave the company detailed entitlements and obligations. Specifically, TelMex was granted a six-year monopoly over long-distance and international public telecommunications services, but was required to offer interconnections to competing service providers when this exclusive franchise expired [4]. TelMex was also required to offer service at tariffed rates, subject to price cap regulation, with all rates set to recover the associated marginal costs and avoid cross-subsidization [5]. Finally, the Title of Concession subjected TelMex to stringent network expansion and quality-of-service goals, including 12% annual growth in access lines.

The cellular telephone market in Mexico, meanwhile, has been substantially liberalized since 1990. Mexico City, in particular, has eight licensed cellular service providers—a far more competitive market than exists in most of the world.

While most noncellular telephone service in Mexico is provided by TelMex, a number of services are also provided by SCT Telecom, a government carrier formed in 1990 and owned by the Secretariat for Communications and Transportation (SCT). SCT Telecom offers long-distance and local service to rural areas, and fax, telex, and public switched data service. For the most part, TelMex and SCT Telecom do not compete directly with one another [6]. Until 1995, the legal framework for telecommunications service in Mexico consisted of provisions of the 1917 Constitution, a telecommunications act passed in 1939, and the Title of Concession granted by the government to TelMex. In June of 1995, however, President Ernesto Zedillo signed a new telecommunications law establishing a framework for long-distance and international competition by 1997. Under the new law, the SCT may grant concessions to competing long-distance carriers and TelMex must negotiate reasonable interconnection arrangements with those carriers [7]. New carriers may commence operation when the TelMex monopoly expires in August 1996.

The new law also creates competition in the provision of satellite services. SCT Telecom will continue to operate the Morelos (domestic) and Solidaridad (regional) satellite systems and provide access to INTELSAT, but it will compete with other entities that are granted concessions by the SCT to transmit signals to and from foreign satellite systems or to use geostationary orbital slots to serve Mexican subscribers.

Finally, under the new law, SCT Telecom will remain the sole provider of telegraph and radiotelegraph services.

Perhaps the most significant effect of the new law will be its impact on the profitability of TelMex. Since privatization, Mexico has had the most profitable public network of all OECD nations—a level of performance that has helped TelMex meet the ambitious universal service obligations set out in its concession. The future will tell whether this level of performance can be sustained in the face of competition.

11.3 THE UNITED STATES

The provision of telecommunications in the United States has two historic features that distinguish it from the telecommunications regimes of most nations. The first of these is the decision to place telecommunications in the hands of regulated, private corporations, rather than government administrations. The second is the division of regulatory oversight of telecommunications between the central government and the states. These two features shaped the evolution of regulation in the telecommunications industry of the United States, and they are now shaping the course of deregulation.

11.3.1 Regulation by the Federal Government

The central government of the United States regulates telecommunications principally through the FCC, which was created by Congress in the Communications Act of 1934, and through decisions issued by federal courts under the antitrust laws. We review each of these sources of regulation in turn.

Regulation by the Federal Communications Commission

The Communications Act gives the FCC jurisdiction over "all interstate and foreign communication by wire or radio...which originates and/or is received within the United States..." [8]—a jurisdiction that includes radio and television broadcasting and cable television, as well as telephone, telegraph, and other telecommunications services [9]. As the following sections explain, the FCC applies this broad jurisdiction very differently to different classes of services and service providers.

FCC Regulation of Common Carriers

Title II of the Communications Act [10] defines the FCC's jurisdiction over so-called common carrier services—a category that includes all traditional telephone and telegraph services and also extends to newer technologies that are offered to the public at large on a nondiscriminatory basis. Providers of services regulated under Title II must accept significant limitations on their ability to make ordinary business decisions, including pricing, investment in new facilities, and the terms and conditions of interconnection with other carriers. We review some of these constraints in the following sections.

Restraints on Common Carrier Pricing and Investment in Facilities

According to Section 203 of the Communications Act [11], all carriers must publish a schedule of their rates—called a *tariff*—that may be viewed and challenged by anyone who believes the rates contained in the tariff to be unreasonable or unreasonably discriminatory. (Such a challenge can delay the effectiveness of a tariff and may result in a finding that the tariffed rates are unlawful.) To the extent common carriers must comply with the tariff process, their ability to change prices in the face of competition or changing customer needs is limited, as is their ability to conceal their pricing decisions from their competitors.

Besides enforcing the Act's general requirements that common carrier rates be published, reasonable, and nondiscriminatory, the FCC by regulation has exercised detailed oversight of carriers' earnings. In recent years, the FCC has begun to relax its control over earnings in response to growing competition in many telecommunications markets.

The traditional form of price regulation is called *rate-of-return*, or *earnings*, regulation, which the FCC still applies to interstate rates of the smaller, local exchange carriers. Under this system the FCC does not set the carriers' rates directly, but permits carriers to recover their reasonable expenses incurred in providing regulated service, plus a rate of return on investment equivalent to the carriers' cost of capital. Once this number—called the *revenue requirement*—is ascertained, the carrier may set its rates at a level that will yield the revenue requirement at the anticipated level of demand. Under rate of return, the carriers' discretion to price particular services is limited only by the Act's prohibition against rates that are unreasonable or unreasonably discriminatory [12].

Beginning in the late 1980s, the FCC has begun to replace rate-of-return regulation with a system known as *incentive*, or *price cap*, regulation. Under this approach the FCC sets the carrier's rates and permits earnings to vary according to the carrier's ability to reduce its cost of service. The FCC adjusts the maximum rates periodically to reflect industrywide efficiency gains offset by inflation, and requires earnings above a certain level to be "shared" with ratepayers. At this writing, price cap regulation is mandatory only for the regulated interstate services of carriers with annual revenues from regulated services of $100 million or more (called Tier 1 carriers).

The FCC's traditional regulation of common carriers extends beyond scrutiny of rates and earnings. Notably, Section 214 of the Act requires common carriers to seek the FCC's permission before they build, extend, or abandon the facilities through which they provide service to the public. Such permission may be granted only where the FCC finds that the proposed investment, extension, or abandonment of service is in the public interest.

The Section 214 process was an important regulatory tool in the days when all common carrier services were provided by monopolists. Because rate-of-return regulation effectively ensured that carriers would recover the costs they incurred, telephone companies lacked strong market incentives to invest prudently. Regulatory oversight was therefore needed to protect ratepayers from bearing the cost of unnecessary or needlessly expensive facilities. Similarly, the lack of alternative service providers called for a high burden of justification on carriers that proposed to discontinue service to a community.

After the FCC and the antitrust courts permitted new competitors to enter particular telecommunications markets, some of the Communications Act's restrictions on the business decisions of common carriers appeared to the FCC to restrict the ability of carriers to respond to competition and meet the changing needs of their customers. The result was a series of FCC decisions that streamlined or eliminated these requirements for carriers the FCC classified as *nondominant*—a category that came to include everyone except AT&T and the local telephone companies [13].

Unfortunately for the FCC, the courts did not agree that the Communications Act granted the FCC the latitude to eliminate requirements that the Act plainly imposed on all common carriers. In particular, the United States Supreme Court ruled that the FCC could not eliminate the tariffing requirements of Section 203 of the Act [14], and a lower federal appellate court has ruled that the FCC may not even permit carriers to file tariffs containing a "range of rates" rather than specific prices [15].

The courts have not been asked to decide whether the FCC's elimination of Section 214 requirements for nondominant carriers is lawful under the Act. For the moment, therefore, Section 214 applies only to dominant carriers and is applied principally (as we saw in Chapter 7) to carriers of international services subject to the FCC's jurisdiction.

Regulation of Interconnection

Among the FCC's most challenging tasks is the management of competition in the U.S. telecommunications industry. As various monopoly markets have opened to new entry, the FCC has developed rules to ensure that the local exchange carriers—and the Bell companies in particular—do not use their control over the local telephone networks to impose disadvantages on their competitors.

A number of approaches to this problem have been tried. The most drastic might be called the *quarantine* approach, which simply prohibits participation by certain carriers in certain markets. Examples of the quarantine are the cable–telephone company cross-ownership restriction (about which we say more in a moment) and the court-ordered prohibition against Bell company participation in the long-distance and manufacturing markets [16]. An intermediate approach requires carriers to offer certain services through separate subsidiaries (the basis, for example, on which the Bell companies are permitted to offer cellular telephone service). A third approach—and the most difficult to administer—permits carriers to offer competing services, but imposes accounting and technical requirements to ensure that those carriers treat competitors fairly. This is often called the *nonstructural safeguards* approach, and is the source of some of the most complex rules the FCC has enacted.

A few of the FCC's more notable nonstructural safeguards rules are worth reviewing. Specifically, telecommunications markets in the United States have been affected significantly by *equal access*, the *Computer III* rules, and the FCC's rules concerning customer premises equipment (CPE).

Equal Access

When the FCC began to permit competition in the long-distance market, most long-distance services were still provided by AT&T and its wholly owned Bell System companies. The FCC was called upon, therefore, to supervise the arrangements by which the Bell companies provided their long-distance competitors with interconnection to the local telephone networks [17]. Later, when a federal court ordered AT&T to divest the Bell companies, the FCC developed rules to govern access in the postdivestiture environment.

Put very generally, the equal access process gives the Bell companies obligations to both the long-distance carriers and their customers. For the long-distance carriers, telephone companies (Bell and independent alike) must provide interconnections, upon reasonable demand, that are equivalent in type and quality to those the companies provide to AT&T's long-distance operations [18]. Access of this kind includes the ability of each long-distance company's customers to reach the carrier by dialing the same number of digits. For the long-distance carriers' customers, the telephone companies must provide an opportunity to presubscribe to a carrier of choice. Where equal access first is initiated in a service area, the local company must circulate ballots and give customers an opportunity to choose a long-distance carrier [19]. Once equal access has been implemented in an area, the telephone company is responsible for prompt processing of customer requests to change their carriers [20].

In addition to the detailed rules governing access for long-distance carriers, the FCC has also enacted rules governing access for mobile radio carriers and competing providers of long-distance access services [21].

The Computer III Rules

Some of the FCC's most complex proceedings have involved the terms under which the Bell companies may provide enhanced services—that is, voice mail, remote data processing, and other services that manipulate, rather than merely transmit, customer-provided information.

At one time, the FCC required the Bell companies to offer enhanced services only through separate subsidiaries. The FCC later determined, however, that the separate subsidiary requirement was inefficient and unnecessary. The result was a series of rules intended to ensure that the Bell companies would offer such services without imposing disadvantages on other enhanced-service providers who also needed access to the Bell companies' local networks.

One element of these *Computer III* regulations is the requirement that the Bell companies file Open Network Architecture (ONA) plans, which are menus of facilities and services through which enhanced-service providers (including the Bell companies' own enhanced services) may obtain flexible access to the local networks [22]. Any changes to the services offered in an ONA plan require prior notice to the industry, and the Bell companies must answer requests for new services within 120 days of their receipt.

When a Bell company proposes to offer an enhanced service, it must file with the FCC a Comparably Efficient Interconnection (CEI) plan, demonstrating to the FCC that it will offer to competing providers of the same service interconnections that are comparable to those the Bell company will provide to itself. The CEI plan need not show that interconnections will be precisely equal, but must demonstrate that they will permit roughly comparable service quality to customers and will impose no competitive disadvantage on the non-Bell provider.

The *Computer III* rules also include specific accounting safeguards, network disclosure rules, and rules for the protection of customer privacy. Specifically, Bell companies must disclose changes in their networks that may affect the interconnection arrangements of competitors, they must maintain books of account based on cost allocation manuals approved by the FCC to ensure that regulated services are not subsidizing the Bell companies' enhanced services, and they must not use information provided by individual and small-business customers in connection with telephone service to market the Bell companies' enhanced services if the customers have directed that the information not be used in this way.

Customer Premises Equipment

The Bell companies are permitted to provide (but not manufacture) telephones and other equipment used on the premises of their customers in connection with telecommunications services. In order to ensure that the Bell companies do not use their control of the local network to the detriment of competing providers of CPE, the FCC has enacted rules similar to those of the *Computer III* regime [23]. Those regulations include network disclosure rules, rules to protect customer information, regulations governing network access, a requirement that the Bell companies permit competing CPE providers to market their products jointly with Bell network services, and accounting rules intended to prevent cross-subsidization of Bell-provided CPE.

FCC Regulation of Radio Services

The Communications Act authorizes the FCC to regulate nongovernmental uses of the RF spectrum in the United States [24]. The FCC's radio regulation consists, broadly, of two functions: (1) the allocation of the available radio frequencies among various uses (e.g., broadcasting, common carrier, private radio, marine, and aeronautical), and (2) the licensing of persons to use radio equipment for these prescribed uses.

The allocation function has grown more complex as the number of uses to which radio can be put has grown. When the technology was first invented, it was used primarily for maritime and military applications. With few users competing for the available frequencies, persons wishing to operate transmitters could choose any frequency that suited their needs. A system of registration with the U.S. Department of Commerce was sufficient to keep track of RF usage.

As time went on, however, radio proved itself adaptable to a wide range of applications. Most dramatically, the growth of broadcasting in the 1920s created enormous new demand for frequencies. To radio broadcasting, within a handful of decades, were added television broadcasting, long-distance telephone service over microwave radio, communications between satellites and Earth stations, mobile radiotelephone service (including cellular telephone), personal communications service, and other applications. All of these competing demands have had to be accommodated within an electromagnetic spectrum whose limits are set by nature [25]. In deciding how segments of the electromagnetic spectrum will be assigned to various applications, the FCC has had to take a number of factors into account. Notably, some radio transmissions (such as satellite signals) cross national boundaries and must be made on frequencies assigned by international agreement. Even where international constraints are not present, the FCC must still assign spectrum according to the physical characteristics of the competing services [26] and the probable demand for each service [27]. As the spectrum has filled with users, allocation of spectrum for new services may require that existing users be moved to other frequencies [28].

The FCC's licensing function, which carries out the expansive mandate of the Communications Act to manage the resource in the "public convenience, interest, or necessity," has been the source of the most detailed regulation of radio operators. Some of those regulations are technical, while others relate exclusively to broadcast stations and dictate, in limited ways, the content of the programming carried by the licensee.

Technical licensing requirements, while expressed in engineering terms, may reflect important policy choices. In determining the power output of broadcast stations, for example, the FCC strikes a balance between having a few powerful stations on each frequency, each sending its signal over a large area, or having many lower power stations serving smaller markets. If the power limits are set too low, the size of the market may be insufficient for stations to survive economically; if the limits are set too high, the number of programming sources available to the public may be needlessly small [29].

Limitations on the content of broadcast programming have varied according to the makeup of the FCC and evolving notions of the public interest. The resulting regulations have been as varied as a ban on broadcast cigarette advertising, mandatory minimums of quality children's programming, regulations intended to ensure access to the airwaves by political candidates, and limitations on indecent speech. The courts have tended to uphold these limitations (which would be unconstitutional if imposed on newspapers or other organs of mass communication) on the grounds that the electromagnetic spectrum is a scarce national resource that must be managed in the public interest, and that broadcasting is somehow more pervasive and intrusive than print media and calls for especially vigilant regulation [30].

Perhaps the most dramatic recent development in the FCC's licensing procedures is the use of auctions to decide who will receive the first set of personal communications service (PCS) licenses. Through most of its history, the FCC has assigned licenses either on a first-come, first-served basis, through proceedings to determine which applicants best would serve the public interest [31], or through lotteries. Only with the passage by Congress of the Omnibus Budget Reconciliation Act of 1993, however, did the FCC acquire the authority to sell licenses through an auction process. The use of this procedure for PCS licensing brought billions of dollars to the U.S. Treasury.

Before leaving the subject of the FCC's regulation of radio, we should note the difference in purpose and approach between Title III regulation and common carrier regulation under Title II. Title II regulation is intended primarily to ensure the widespread availability of telephone service and to prevent common carriers, who traditionally have enjoyed protected monopolies, from exploiting their customers. Consistent with these objectives, the focus of Title II regulation has been on oversight of the earnings, pricing, and investment decisions of common carriers. Title III regulation, on the other hand, is intended not to control monopolists, but to allocate a scarce resource—the electromagnetic spectrum—and ensure that it is used in the best interests of the public. Title III regulation leaves commercial users of the radio spectrum with almost complete freedom to price their services and invest in facilities [32], but imposes public-interest limitations on their use of the spectrum assigned to them.

Title VI: FCC Regulation of Cable Television

The invention and spread of cable television presented a challenge to the FCC's ingenuity. The service did not exist when the Communications Act was passed and did not appear to fit easily into any of the categories established by the Act.

Cable television presented certain features typical of common carriers. It was paid for by subscribers (unlike broadcast radio and television, which were supported by advertisers and transmitted to the public free of charge), and it was offered by only one service provider in each service area, therefore raising the possibility of monopolistic abuse of customers. But it also carried television programming, some of it originating with broadcasters and all of it resembling the programming offered by broadcast stations; and it seemed to many that the medium should be regulated under the broadcast scheme the FCC had developed under Title III of the Act.

At first, the FCC found no basis to exercise jurisdiction over cable television under either Title II or Title III of the Communications Act [33]. Cable was therefore regulated primarily by states and municipalities, which had jurisdiction because of the cable companies' need to run their facilities along public rights of way.

Beginning in the 1960s, the FCC reversed its earlier determination that cable television did not fit any basis of FCC jurisdiction. In a series of decisions, the FCC moved to protect local broadcasting from cable competition by requiring the cable operators to carry all broadcast programming within the cable systems' service areas (*must-carry* rules), and by requiring cable operators to avoid carriage of distant broadcast signals that duplicated locally available programming [34]. The U.S. Supreme Court upheld these rules as "reasonably ancillary" to the FCC's jurisdiction over broadcast television [35].

Eventually, however, the FCC's efforts to improvise a basis for jurisdiction over cable TV were rebuffed. In *FCC v. Midwest Video Corp.* [36], the U.S. Supreme Court declared that the FCC exceeded its authority when it required cable operators to provide access to their facilities for certain classes of local programmers. (Unlike the must-carry and nonduplication rules, these regulations were not intended to protect broadcasters, but rather to give certain favored groups a cable TV outlet.) In the Court's view, the access rules resembled common carrier obligations of the kind the FCC imposes under Title II of the Act, and could not be regarded as ancillary to the FCC's jurisdiction over broadcasting.

Congress rescued the FCC from its dilemma by enacting the Cable Communications Policy Act of 1984 (usually known simply as the Cable Act), which added a new Title VI to the Communications Act and provided for a regime of mixed broadcast and common carrier regulation of the cable television industry. Among other provisions, the Act codifies the access requirements that were struck down in *Midwest Video*, limits the ability of local authorities to regulate cable rates, severely restricts the ownership of cable companies by telephone companies, and reserves the right of states and localities to impose franchise fees and other regulation to the extent not incompatible with the provisions of the Act [37].

Unfortunately, the 1984 Cable Act did not lead to lower prices for consumers. Expected competition from wireless cable and direct broadcast satellite was slow to develop, leaving consumers dependent, for the most part, on monopoly providers of cable services. Congress ordered a General Accounting Office study of the industry, which confirmed that cable rates had risen at three times the consumer price index since passage of the 1984 Act [38]. Based on these disappointing results, Congress decided to reregulate the industry.

The Cable Television Consumer Protection and Competition Act of 1992 (usually referred to as the 1992 Cable Act) [39] provides that all cable systems not subject to "effective competition" are subject to rate regulation [40]. (The mechanics of the rate regulation scheme are complex and apply only to the "basic tier" of cable services.) Unfortunately, the ratemaking provisions of the 1992 Act are widely regarded as a well-meaning failure, with cable operators exploiting loopholes in the Act and the FCC's enabling rules to avoid the decreases the Congress hoped to achieve.

Other aspects of cable regulation have been challenged in the courts with some success. Notably, both the must-carry rules and the restriction on telephone company ownership of cable systems have been rejected by the courts as contrary to the right of free expression guaranteed by the First Amendment to the U.S. Constitution. In the case of the cross-ownership restrictions, the ban has been rejected outright by various appellate courts, and the FCC has abandoned its enforcement of the ban. In the case of must-carry, the U.S. Supreme Court ruled that Congress had not made adequate findings as to the necessity for the requirement [41].

Finally, the rate regulation provisions of the 1992 Cable Act will probably be eliminated by legislation. In light of the failure of the present rules to lower rates and the long-awaited advent of competition from direct broadcast satellite and telephone company video services, the approach of the 1992 Act has lost much of the Congressional support it enjoyed only a few short years ago.

Regulation by the Antitrust Courts

The antitrust laws of the United States have had a profound effect on the evolution of U.S. telecommunications industry. In order to appreciate how important these laws have been, it is necessary to review some history.

During the early years of telephony in the United States, local service was provided in many localities by competing companies. Some observers perceived this process as wasteful and contrary to the public interest, and in 1913 an agreement between the president of AT&T and the Attorney General of the United States (known to history as the Kingsbury Commitment) created an official condition of monopoly in telecommunications in the United States. After the Kingsbury Commitment, most telephone subscribers bought their service (both local and long-distance) from AT&T and its wholly owned Bell subsidiaries. Where other companies provided service (chiefly in rural areas and smaller cities), they did so as local monopolists with whom AT&T had pledged not to interfere. The Bell companies bought their switches, cables, and other equipment from a wholly owned manufacturing subsidiary (Western Electric), which also made the telephones and other items of CPE that were leased to Bell's subscribers.

In exchange for their monopoly, telephone companies in the U.S. could earn no more than the regulatory commissions permitted them to earn. The FCC limited the earnings they could realize from interstate service, and the states limited the earnings they could realize from intrastate service.

During the quiet decades of regulated monopoly, the antitrust laws had little effect on the Bell companies. Since competition was not allowed in the telephone business, the antitrust laws (which are designed to protect the competitive process from monopolistic or collusive interference) had no role to play.

This began to change, however, when the FCC began to allow limited competition in certain markets—notably, in interstate calling and manufacturing of customer premises equipment. Once competition was permitted, the Department of Justice and the FCC both ascertained that the Bell companies were imposing disadvantages on their new competitors. Governmental scrutiny of the Bell companies' conduct culminated in the 1974 filing of a massive antitrust suit against AT&T [42].

The government's lawsuit was based primarily on a judge-made rule of antitrust law called the *essential facilities doctrine*. Under this doctrine, a firm that controls a facility that cannot practicably be duplicated, and access to which is essential in order for other firms to compete in a market, must make that facility available to its competitors on reasonable terms and conditions [43]. The government charged that the Bell companies controlled the local telephone networks, which could not practicably be duplicated, and that the Bell companies' competitors (particularly in the long-distance market and the market for sales of telephone equipment to customers) could not compete without reasonable access to those local networks. The government further charged that the Bell companies had provided their competitors with inadequate forms of access to the local networks for the purpose of preserving AT&T's monopoly of the long-distance and customer equipment markets.

The government also charged AT&T with other violations of the antitrust laws. Notably, the lawsuit claimed that AT&T had engaged in below-cost pricing of services for which it faced competition and had abused a "buyer's monopoly" over the market for equipment sold to telephone companies.

Eventually, the government's lawsuit was settled by agreement between the U.S. Department of Justice and AT&T. As a result of that agreement, AT&T was separated from its local Bell companies under the terms of a consent decree entered by the court. AT&T kept its manufacturing and long-distance operations, and the local exchange services of the Bell System were allocated to seven regional holding companies (RHC). The RHCs, in turn, were forbidden to manufacture telecommunications equipment, to carry telephone calls beyond the borders of their service areas, or to provide information services (i.e., remote data processing or other services that manipulated the information transmitted by the customer) [44].

In order to implement the agreement, the territory of the United States was divided into some 160 local administrative and transport areas (LATA). The RHCs were permitted to provide telephone service within the LATAs, but could not transport calls between points in different LATAs.

Implementation of the agreement also required ongoing supervision by the federal judge who had presided over the government's lawsuit against AT&T. For this purpose, the consent decree created a procedure under which the court could modify the agreement from time to time to permit the RHCs to engage in particular manufacturing, long-distance, and information service activities otherwise forbidden to them. If a request to engage in such a service is not opposed by any party to the decree (i.e., the government or AT&T), then the court will grant the request unless it is clearly contrary to the public interest. If the request is opposed by any party, then the court will grant it only if it finds no substantial possibility that the RHC will impede competition in the market it seeks to enter.

This waiver process has resulted in some amelioration of the decree's effect on the RHCs. Most notably, the RHCs have been relieved by waiver of the information services restriction, and are now prevented only from manufacturing telecommunications equipment and offering inter-LATA telecommunications services.

Over the years that have passed since the AT&T consent decree became effective in 1984, the telephone companies' monopoly of local exchange service—the condition that led both to the government's lawsuit and the decree's restrictions—has been eroded by technology and competition. New telephone companies have entered the larger business centers to compete with the RHCs for corporate customers, and cellular telephone and wireless services of all kinds have provided a practical alternative to the copper-wire local networks of the telephone companies. Under these circumstances, there is less reason to fear that the Bell companies, if they are allowed to manufacture equipment and offer inter-LATA service, will exploit their local networks to the detriment of their competitors. For this reason, the Bell companies have gained considerable support for Congressional legislation that eventually will eliminate the restrictions of the consent decree altogether.

Even if the AT&T consent decree is eliminated, the antitrust laws will continue to exercise considerable discipline over the conduct of telephone companies, cable television companies, and other players who control access to customers their competitors want to reach. That discipline will be exercised primarily by litigation (or the threat of litigation) brought either by the government or by aggrieved private parties.

11.3.2 Regulation by the States

The Constitution of the United States limits the powers of the central government and reserves substantial authority to the 50 separate states. This division of powers is expressly upheld in the federal Communications Act, which denies the FCC any regulatory power over "charges, classifications, practices, services, facilities, or regulations for or in connection with intrastate communication service by wire or radio" [45]. Telecommunications services, of course, are not confined within state boundaries and are therefore difficult to divide into intrastate and interstate components for purposes of regulation. The definition of these limits (a subject not coherently addressed in the Communications Act) has been one of the most perplexing tasks of regulation in the United States.

The best guidance on this subject has come from the U.S. Supreme Court, which decided in 1986 that the FCC may not "preempt" state regulation unless two tests are met [46]. First, the FCC may preempt state regulation only where necessary to achieve some valid goal that is within the FCC's jurisdiction under the Act; and second, the FCC may preempt only those aspects of state regulation that cannot be separated into interstate and intrastate components.

In practice, the FCC has tended to regulate the rates charged for service concurrently with the state commissions, and to preempt non–rate regulation on a case-by-case basis. So, for example, the states may impose separate charges for access to telephone company facilities in connection with intrastate cellular telephone calls, and the FCC regulates charges for access to those same facilities in connection with interstate cellular calls. But the states are not allowed to impose their own standards of physical interconnection to facilities that are used to provide both interstate and intrastate cellular calls [47].

The FCC and the state commissions have also worked to avoid conflict in their separate regulatory schemes. In matters of joint regulatory concern, the Communications Act authorizes the formation of joint boards staffed by both state and federal regulators [48].

The states have historically exercised close control over the earnings derived by telephone companies from their intrastate services, relying primarily on rate-of-return approaches. It is fair to say that the chief concern of the state commissions has been to keep basic residential telephone rates as low as possible by imposing above-cost rates for business and optional services and requiring the telephone companies to average their rates geographically by permitting the rates charged to low-cost users (chiefly urban customers) to subsidize the rates charged to higher cost (chiefly rural) customers. The states have also been slower than the FCC to permit new entry into the telephone markets under their control.

In recent years, many state commissions have acted both to permit competition in the provision of local service and to experiment with various types of regulatory reform. Many of the reform programs include price cap rather than rate-of-return oversight for the services of the larger carriers, and a somewhat larger number of states permit at least some leeway in the pricing of services subject to competition.

As the U.S. Congress considers legislation to deregulate the telecommunications industry, new questions of preemption have been raised. Notably, Congress is considering whether to require the states to eliminate rate-of-return regulation and limitations on new competition in the provision of local service. The state regulators, of course, are arguing that their discretion in these areas should be preserved.

11.3.3 The Future of Telecommunications in the United States

The clear direction of regulation in the United States, both at the state and federal levels, is toward less government control and more competition. This process is going forward on many fronts.

One of the most important obstacles to open competition, and one that is under concerted assault, is the AT&T consent decree and its restrictions on the businesses the RHCs may enter. The Bell companies have attacked the restrictions through the waiver process and hope to eliminate it altogether through legislation.

Another perceived obstacle is the system of subsidies that continue to infect rates for telephone service. Some of these subsidies—notably, the Universal Service Fund used to support local rates charged by carriers in rural and other high-cost areas—are explicit. Others—such as geographic rate averaging and higher rates for business than for residential service—are more difficult to quantify. To the extent these subsidies continue to exist, however, they are believed to inhibit competition in the markets for subsidized services. It is likely that the present system of subsidies will be modified substantially by amendment of the Communications Act.

While most talk of telecommunications reform in the United States tends toward the direction of deregulation, some argue that the role of regulation should increase in order to ensure that broadband and interactive services reach as many U.S. households as possible. These "industrial policy" approaches have some support in the White House and Congress, and may result in limited measures to ensure that broadband services reach schools, hospitals, and other users that present particular public-interest concerns.

Notes

[1] For a time this association was known as Telecom Canada. Telesat, the domestic satellite carrier owned partly by the central government and the various domestic carriers, is also a member of Stentor.

[2] For an excellent chronology of liberalization in Canada before the 1993 Act, see W. Melody and P. Anderson, "Telecommunication Reform in Canada," in *Global Telecommunications Policies: The Challenge of Change*, M. Jussawalla, ed., Westport, CT: Greenwood Press, 1993,

pp. 100–101. For more recent developments, see Organization for Economic Cooperation and Development, *Communications Outlook 1995*, pp. 123–125.

[3] TelMex had been formed from two foreign-owned companies, controlled by ITT and Ericsson, in 1947. Control passed to Mexican owners in 1957, and in 1972 51% of the shares were purchased by the Mexican government, which achieved a fourfold increase in access lines (1.1 million to 4.4 million) between 1972 and 1988. With the economic troubles of the mid to late 1980s, the government could no longer sustain this rate of growth and privatization became more attractive.

[4] Technically, the local service market has been open to competition since 1990, but the lack of interconnection arrangements and rules has delayed the entry of new providers into the market. See, for example, Organization for Economic Cooperation and Development [2], p. 117.

[5] Tariffs and supporting material must be submitted by TelMex every four years for review by the Secretariat of Transportation and Communications (SCT). If the SCT and TelMex are unable to agree on the tariffed rates, three experts (one appointed by TelMex, one appointed by SCT, and one agreeable to both parties) may be called upon to resolve the deadlock.

[6] A peculiarity of Mexico's legal environment is that SCT's monopolies of telegraph and broadcasting service are guaranteed by the Constitution. See R. Baica, "Privatization, Deregulation and Beyond: Trends in Telecommunications in Some Latin American Countries," in *Global Telecommunications Policies: The Challenge of Change*, M. Jussawalla, ed., Westport, CT: Greenwood Press, 1993, p. 146.

[7] Under rules to be issued by the SCT, as of 1997 TelMex and its competitors will engage in a balloting process under which telephone customers will presubscribe to the long-distance carriers of their choice. All subscribers will reach their presubscribed carriers by dialing a two-digit access code.

[8] 47 U.S.C. section 152.

[9] Ibid., section 152(a).

[10] 47 U.S.C. section 201 et seq.

[11] 47 U.S.C. section 203.

[12] For a more complete explanation of rate and earnings regulation, see C. Kennedy, *An Introduction to U.S. Telecommunications Law*, Norwood, MA: Artech House, 1994, pp. 5–29.

[13] See, for example, *Policy and Rules Concerning Rates for Competitive Common Carrier Services and Facilities Authorizations Therefor*, 85 FCC 2d 1 (First Report and Order released November 28, 1980).

[14] *MCI v. AT&T*, 114 S.Ct. 2223 (1994).

[15] *Southwestern Bell Corp. et al., v. FCC*, 43 F.3d 1515 (D.C. Cir. 1995).

[16] See pp.189–191.

[17] We discuss on pp.189–191 the antitrust problems posed by interconnection.

[18] *Investigation Into the Quality of Equal Access Services*, 60 Rad. Reg. 2d (P&F) 417 (Memorandum Opinion and Order released May 23, 1986). This requirement is also articulated in the AT&T antitrust consent decree, about which we say more later in this chapter. *United States v. American Telephone and Telegraph Co.*, 552 F. Supp. 226 (D.D.C. 1982).

[19] *Investigation of Access and Divestiture Related Tariffs*, 101 FCC 2d 911 (Memorandum Opinion and Order released June 12, 1985).

[20] *Policies and Rules Concerning Changing Long Distance Carriers*, 7 FCC Rcd 1038 (Report and Order released January 9, 1992).

[21] *In the Matter of the Need to Promote Competition and Efficient Use of Spectrum for Radio Common Carrier Services*, 2 FCC Rcd. 2910 (Declaratory Ruling released May 18, 1987); and *In the Matter of Expanded Interconnection With Local Telephone Facilities*, 6 FCC Rcd 3259 (Notice of Proposed Rulemaking and Inquiry released June 6, 1991).

[22] *Filing and Review of Open Network Architecture Plans*, 5 FCC Rcd 3103 (Memorandum Opinion and Order released May 8, 1990).

[23] *Furnishing of Customer Premises Equipment by the Bell Operating Telephone Companies and the Independent Telephone Companies*, 2 FCC Rcd 143 (Report and Order released January 12, 1987).

[24] The U.S. government reserves for its own use roughly half the available frequency spectrum. These uses are regulated not by the FCC, but by the National Telecommunications and Information Administration.

[25] For a scientific explanation of spectrum scarcity, see Jackson, "The Allocation of the Radio Spectrum," *Scientific American*, February 1980, p. 34.

[26] Microwave, for example, lends itself better to point-to-point applications than it does to broadcast transmissions.

[27] The FCC's allocation of spectrum for personal communication service (PCS), for example, was controversial because some commenters (and one commissioner) felt that some licensees were given too little spectrum (20 MHz) to become economically viable.

[28] In order to find spectrum for PCS services, for example, the FCC decided to displace some existing users of private microwave service.

[29] The FCC faces similar choices in licensing certain nonbroadcast applications. In the PCS proceeding, for example, the FCC weighed the extent of the geographic area within PCS licenses should be awarded. (The FCC even considered issuing some number of nationwide licenses—a course it elected not to take.)

[30] For an excellent critical discussion of the various rationales for broadcast regulation, see T. Krattenmaker and L. Powe, Jr., *Regulating Broadcast Programming*, Cambridge, MA: MIT Press, 1994. See also *National Broadcasting Co. v. United States*, 319 U.S. 190, 210 (1943).

[31] Renewals of broadcast licenses, which are assigned for a fixed term, are also decided between incumbents and competing applicants according to which operator will best serve the public interest.

[32] Common carriers that happen to use radio to provide their services, of course, are subject to both Title II and Title III regulation.

[33] *Frontier Broadcasting Co. v. Collier*, 24 FCC 2d 251 (Memorandum Opinion and Order 1958).

[34] *Application of Carter Mountain Transmission Corp.*, 32 FCC 459 (1962), *aff'd*, 321 F.2d 359 (D.C. Cir.), *cert. denied*, 375 U.S. 951 (1963); *Rules re Microwave-Served CATV*, 38 FCC 683 (First Report and Order 1965), *aff'd sub nom., Black Hills Video Corp. v. FCC*, 399 F.2d 65 (8th Cir. 1968); *Rules re Microwave-Served CATV*, 2 FCC 2d 725 (Second Report and Order 1966), *aff'd, Black Hills Video Corp. v. FCC*, 399 F.2d 65 (8th Cir. 1968).

[35] *United States v. Southwestern Cable Co.*, 392 U.S. 157 (1968).

[36] 440 U.S. 689 (1979).

[37] While the 1984 Act solidified the FCC's jurisdiction over cable television, its overall aim was to relax regulation of the industry in the hope that growing competition would lead to lower rates and increased alternatives for consumers. See R. G. Prohias, "Longer Than the Old Testament, More Confusing Than the Tax Code: An Analysis of the 1992 Cable Act," *CommLaw Conspectus*, Vol. 2, 1994, pp. 81–83.

[38] H.R. Rep. No. 862, 102d Cong., 2d Sess. 55 (1992).

[39] Pub. L. No. 102-385, 106 Stat. 1460, codified at 47 U.S.C. section 521-555 (1992).

[40] The burden is on the cable operator to show that it is subject to effective competition.

[41] Turner Broadcasting System v. FCC, 129 L.Ed.2d 497 (1994).

[42] For a detailed discussion of the Government's case, see *United States v. AT&T*, 524 F. Supp. 1336 (D.D.C. 1981).

[43] See *Otter Tail Power Co. v. United States*, 410 U.S. 366 (1973); and *United States v. Terminal Railroad Association of St. Louis*, 224 U.S. 383 (1912).

[44] See *United States v. American Telephone and Telegraph Company* and *United States v. Western Electric Company*, 552 F. Supp. 226 (D.D.C. 1982).

[45] 47 U.S.C. Section 152(b).

[46] *Louisiana Public Service Commission v. FCC*, 476 U.S. 355 (1986).

[47] The FCC found that the same physical facilities are used for both interstate and intrastate calling and cannot be divided between the two jurisdictions, and that restrictive state regulation of interconnection could frustrate FCC policy by denying cellular carriers access to the interstate network. Preemption of the physical terms of access, therefore, was justified. 2 FCC Rcd. 2910 in the Matter of the Need to Promote Competition and Efficient Use of Spectrum for Radio Common Carrier Services (1987).

[48] The state members of the joint boards, however, have no voting power. See 47 U.S.C. section 410(c).

12 Telecommunications and the European Union

The European Union is the world's largest telecommunications market. Its 15 members have a combined population of over 400 million inhabitants. As a market, it is of particular interest to all telecommunications providers not only because of its size, but because its high standard of living and highly educated populace drive the demand for both business and residential advanced telecommunications services. The share of telecommunications in the EU's gross national product is expected to more than double in the next five years [1]. The EU has recognized these facts and has declared that a liberal telecommunications policy is an essential component of economic growth and unity. Unlike the United States, Europe has a long history of internal struggle for national and geographic definition. None of the wars that afflicted the continent and decimated its population came as a surprise, however. For the past 150 years, since the invention of the telegraph, the various nations of Europe have considered government control of their telecommunications networks essential to national survival. In this context, comprehensive coverage prevailed over economic operation of the network, and the government-controlled telecommunications entities, the PTTs (for the French *Poste, Télégraphe, Téléphone* [2]), were heavily subsidized in order to extend service to sparsely populated and remote geographical areas of strategic importance, such as the French Ardennes mountains and areas on the North Atlantic Coast of Ireland. One of the military and strategic goals of this extensive communications network was to facilitate general mobilization and military communications in times of war. The military importance of communications also explains the reluctance of European countries to privatize their networks [3] and the insistence that they not be shared with or controlled by foreigners. One of the main concerns of the European nations when the ITU was formed at the beginning of the century to promote compatibility and interconnection was that nations retain the right to suspend international communications in case of armed conflict. Today, the political climate of European unity and the advances in telecommunications technology have rendered this approach obsolete. Yet the EU is still struggling with concepts and approaches to this area that are taken for granted in the United States.

12.1 A BRIEF HISTORY OF THE EUROPEAN UNION

The history of Western Europe is the tale of the bloody wars that shaped it. The EU is the direct result of the last of those wars, and its origins center around its main three participants, the United Kingdom, Germany, and France. Germany and France had been traditional enemies, but the intensity and scope of their territorial disputes increased with the unification of Germany in the 1870s. As a result of that struggle, the two countries suffered through three wars in 80 years: the Franco-Prussian war of 1870, the Great War of 1914 (World War I), and the Second World War. After the last of these conflagrations, politicians on both sides realized that the best way to prevent another devastating war would be to integrate the economies of those two countries and of the other European participants in World War II in a way that would make another war economically impossible. The first step was the creation of the European Coal and Steel Community (ECSC) by the Treaty of Paris in 1951 [4]. The Treaty of Paris merely established a common market for coal and steel, but it was an important first step toward integration. ECSC was followed by the European Common Market, composed of the ECSC members France, Germany, Italy, Belgium, the Netherlands, and Luxembourg (those last three countries are collectively known as Benelux), and created by the Treaty of Rome in 1957. The aim of the Treaty of Rome was not only to promote economic integration, but also, as stated in its preamble, "to lay the foundations of an ever closer unity among the peoples of Europe." Great Britain joined after some considerable internal struggle in 1973, along with Denmark and Ireland.

Great Britain was initially reticent on the idea of joining the Common Market for political reasons. As an island, it had been spared the invasion of the Nazi armies and did not have pressing border concerns. Moreover, the French president, General De Gaulle, opposed the accession of a traditional political rival of France to the Treaty of Rome. Finally, Great Britain was more concerned with maintaining its privileged trading relations with the Commonwealth and its former colonies, and resented the loss of national sovereignty that accession to the Common Market would necessarily imply. In the end, the economic success of the Common Market made it impossible for Great Britain to remain at its margins any longer, and it reluctantly joined in 1977. However, the United Kingdom remains a source of dissent within the EU, especially in the telecommunications arena, where its positions are significantly more liberal that those of the other members.

The European Community was subsumed into the EU in 1994. It is now composed of 15 countries: Austria, Belgium, Denmark, France, Finland, Germany, Greece, Ireland, Italy, Luxembourg, the Netherlands, Norway, Portugal,

Spain, and the United Kingdom. Sweden turned down membership in a referendum. Several other countries, including Turkey, are considering joining the EU. The European Community has evolved from being a framework for the removal of trade barriers in coal, steel, oil, natural gas, and other natural resources to becoming an all-encompassing supranational European federal system with authority over all aspects of a country's economy and social policies. This process was achieved by the adoption of a White Paper proposed by the United Kingdom that recommended the creation of an integrated market by 1992. This White Paper became the Single European Act which went into effect in 1987. Among the goals of the Single European Act were the elimination of all custom duties; harmonization of tax rates; free movement of people, goods, and capital; and the creation of a single Europewide telecommunications market. A further step toward unification was taken by the Maastricht Treaty of 1991, in which the adoption of a single European currency, the ECU (for European Currency Unit) was announced for 1999, as was the adoption of a single stand on defense and foreign policy. The initial failure of Denmark to ratify the treaty [5] and the narrow margin by which it was approved in France, together with the onset of the 1992 recession, placed the future of the total integration envisioned by the Maastricht Treaty in question. However, there is little doubt that integration will proceed, and that the goals of the Single European Act will become a reality.

The Treaty of Rome acts as a constitution in that it announces broad principles to be achieved and leaves it up to the institutions it creates to interpret the treaty and implement the laws necessary to achieve its goals. The supreme courts of a number of member states have held that the treaty amends their national constitutions. The principal institutions of the EU created by the treaty are the European Commission, the Council of Ministers, the European Parliament, the Court of Justice, and the Economic and Social Committee. Of these, the European Commission has been by far the most active in the telecommunications area. Its role is to propose new policies, to act as the executive power to implement EU policies and treaties, and to conduct the foreign relations of the EU. Its 20 commissioners are appointed by "common accord of governments of the member States" on the grounds of their technical competence [6]. Once appointed, however, they must remain independent and may not be influenced by their governments. The commission is divided into 19 directorates general (DG) with extensive investigative powers, each covering a substantive area. DG XIII is responsible for telecommunications, but it has been DG IV, the Competition Directorate, which has been responsible for the rapid evolution in this area. Each directorate is headed by a director general and is the responsibility of a given commissioner.

The Council of Ministers is a political body composed of ministers of each member nation. Each country may have only one minister, but the identity of that minister is not fixed. Rather, it depends on the topic being discussed. Thus, when a measure dealing with transportation is being discussed, the Council of

Ministers will be composed of the ministers for transportation of each member country. If the issue shifts to agriculture, the agriculture ministers will replace their colleagues. Nor is the council necessarily composed of ministers. In every case, the highest national authority responsible for a particular topic will be that country's representative in the council. When the heads of state gather to discuss an issue, the Council of Ministers becomes the European Council. The council must approve all legislative measures by qualified majority. Voting is proportional based on the population of the country represented. Thus, France, Germany, the United Kingdom, and Italy have ten votes each; Spain has eight; Belgium, Greece, the Netherlands, and Portugal five; Denmark and Ireland three, and Luxembourg two. However, when "vital national interests" are at stake, a country may insist on unanimous voting [7].

The European Court of Justice is responsible for the interpretation of the treaty and all European legislation. It is the supreme legal authority of the EU, with jurisdiction over constitutional, social, economic, and administrative issues. The European court sees its role as promoting European unity and integration, and it has taken what common law observers would consider an activist role in achieving it. A major difference between the court and common law courts is that the former is not bound by any precedent. However, the court has tried to preserve consistency. Thirteen independent judges sit on the European Court of Justice.

The Economic and Social Committee is a consultative body that provides its opinion and advice concerning issues within the expertise of its members.

The European Commission has been by far the most active of the European institutions in telecommunications matters.

12.2 TELECOMMUNICATIONS: A EUROPEAN PERSPECTIVE

The EU did not turn its attention to telecommunications until the mid-1980s. Until then, all PTTs had remained monopolies under the assumption that this was the best way to ensure the provision of universal service that is their duty as a public utility. Since local calling is considered a public utility, the prices of local calls were kept artificially low by subsidies from the artificially inflated price of long-distance and international calling, regarded as a luxury. At the same time, the telecommunications sector was an important source of revenue for European governments.

Although, as noted above, DG XIII is responsible for telecommunications, it is the much stronger and more powerful DG IV, the Competition Directorate, that has been responsible for dismantling the PTT monopolies through a combination of political will and wile. Because of its profitability and strategic importance, the telecommunications sector had been considered a sacred cash cow by all the European governments, who were not likely to give it up without a fight. The commission had to first firmly establish its power and then wait for the right political climate to make its procompetitive move. With that in mind, the commission undertook telecommunications reform first with careful and conservative steps, and later, with the arrival of Sir Leon Brittan of the United Kingdom as commissioner for competition, with brazen acts and statements. Although Sir Leon's successor, the Belgian Karel Van Miert has been less aggressive in this area, it is unquestionable that the force set in motion by the Brittan Commission is now unstoppable and telecommunications liberalization irrevocable.

It is no coincidence that the real thrust for liberalization came from a British commissioner. The United Kingdom was the first (and so far the only) member state to completely open its domestic market to competition. British Telecom (formerly British Telecommunications) was established as a private company in 1981 by selling the assets of its predecessor, the United Kingdom Post Office. In 1984, the British government gave Mercury Communications, a subsidiary of Cable & Wireless, a license and encouragement to compete with British Telecom. Ironically then, the commission's liberalization process began in 1982 when British Telecom was sued on unfair competition grounds. The *British Telecom* decision was the first case where unfair competition was alleged in a telecommunications context [8]. It involved a transiting arrangement whereby telex messages from other European countries and destined for countries outside of Europe were routed through the United Kingdom in order to take advantage of lower rates and currency fluctuations. British Telecom prohibited private message-forwarding agencies from relaying those messages at the insistence of other European PTTs, who saw this practice cut into their profits. The commission found that British Telecom had abused its dominant position and that competition rules apply to PTT monopolies. In its decision, the commission issued its first warning of things to come by finding that a PTT monopoly must exercise its privileges in accordance with and within the limits of European competition rules. The company was enjoined to lift the prohibition. This case was of particular importance for two reasons: (1) it was the first time that the commission applied competition rules to the telecommunications industry, and (2) British Telecom's actions had followed an ITU recommendation fully supported by other countries aimed at avoiding possible unfair loss of revenue by PTTs. Italy challenged this decision before the European Court of Justice, arguing that the commission's decision was contrary to Article 222 of the treaty, which precludes the EU from meddling in the property ownership regimes of its members. Italy then alleged that Article 222 prevails over Article 86, which prohibits abuse of a dominant position. The European Court of Justice rejected these arguments and upheld the commission's chastising of a national monopoly on competition grounds [9]. Armed with the sanction of the Court of Justice, the commission has relied heavily on the *British Telecom* case to advance its liberalization program. The case established firmly the principle that telecommunications monopolies are subject to the competition rules of the EU treaty, as well as to the articles that guarantee free movement of goods and services across borders. The commission then used the precedent set in the *British Telecom* case to prevent member states from expanding their telecommunications monopolies into new services and technologies and ultimately to dismantle the monopolistic regimes themselves [10].

In 1987, the commission published a Green Paper on Telecommunications [11]. This document went one step further by acknowledging the need for increased competition in telecommunications as a way to provide consumers with "a greater variety of telecommunications services, of better quality and at lower cost" [12]. The commission announced in the Green Paper that the traditional forms of organization in the sector (i.e., state-owned monopolies) were a hindrance to the development of the full potential of new technologies. The Green Paper recognized that "a limited number of reserved services" could be the subject of special monopolistic rights by telecommunications administration in the interest of preserving their public service goals (such as universal service and affordability). However, the commission insisted on a restrictive interpretation of the scope of those reserved services and announced that only basic voice telephony qualified under this exception to the competition goal [13]. The Green Paper did not envision a perennial exemption for telephony. It stated that the monopolistic assumption would have to be revised periodically, and hinted that it may eventually be eliminated. Thus, in the Green Paper, the commission made clear its intention to open the telecommunications sector to competition as far as the political climate would allow at the time. However, instead of guaranteeing the incumbent PTT their existing monopolies, the Green Paper merely recognized their right to "compete alongside other suppliers in the provision of new services" [14]. In a later document commenting on the Green Paper, the commission repeated its intention to push liberalization as far as the political climate would allow then and in the future by declaring that the liberalization process is "iterative, it accepts the existence of a movement, not all aspects of which can be defined today" [15]. With that statement, the commission left the door open for further reforms in the interest of developing "the conditions for the market to provide European users with greater variety of telecommunications services, of better quality and at lower cost"; that is, to fulfill the principle stated in the preamble of the Green Paper. The Council of Telecommunications ministers soon endorsed the goals of the Green Paper in a Council Resolution [16]. To implement its objective, the commission immediately issued two directives, one on telecommunications equipment in 1988 [17] and one on services in 1990 [18].

12.3 THE EUROPEAN COMMISSION COMES OUT OF THE CLOSET

The Green Paper was a clear declaration of principles on the part of the European Commission. Far from deferring to national governments, the commission announced that it would act aggressively to open the telecommunications sector to competition as soon as feasible. The commission was certainly aware that it would encounter strong resistance from member states intent on preserving telephone monopolies. In order to overcome that resistance, the commission took a bold step: the regulations that implemented its new telecommunications policy were adopted under Article 90 of the Treaty Establishing the European Economic Community (EEC TREATY), which arguably permitted the commission to bypass other European Community institutions in promulgating Community laws. The Equipment Directive and the Services Directive were immediately controversial because they were issued under Article 90, which permits the commission to issue directives to member states without consulting the other EU organs, bypassing the council. Because representation in the council is by country, this article is invoked when the commission is seeking to impose legislation that will not be favored by the member states. Directives are addressed to member states and must be implemented through national legislation. They can be issued by the commission or by the council.

Article 90 of the EEC Treaty imposes upon member states the obligation to neither enact nor maintain measures contrary to the treaty in their relationship with state enterprises (known as *undertakings*) and with enterprises to which sites have granted exclusive rights of the same sort. Article 90(3) contains a special enforcement procedure permitting the commission to address directives to the member states directly without first submitting them to the council and the parliament.

The thrust of all of the commission's documents on telecommunications is threefold. Its agenda is to advance liberalization, humanization, and competition in the sector. The directives initially sought to define the line between competitive services and those that can remain monopolies. As time has passed, however, that line has been pushed farther and farther.

The Equipment Directive eliminated all monopolies on telecommunications hardware, while the Services Directive abolished all telecommunications service monopolies except for simple voice, which accounted for 80% of telecommunications services and revenues. In voice, however, the commission exhorted PTTs to align their accounting rates and collection rates with cost. But the most important innovation brought about by the directives is that the commission ordered member states to separate the regulatory from the operational aspects of their telecommunications providers. The two bodies must be independent of each other and maintain an arm's length relationship.

12.3.1 The Equipment Directive

The first step was to liberalize the provision of telecommunications equipment. The commission drew the line at terminal equipment connected to the terminal point of the public switched network (i.e., a telephone or a modem). Although the operators could keep their monopoly over the provision of the network itself, they had to relinquish their special rights to the importation, marketing, manufacturing, or connection of terminal equipment by 1990.

Not surprisingly, recalcitrant member states immediately challenged both directives. France, Italy, Belgium, and Germany [19] appealed the Equipment Directive to the Court of Justice on two grounds: that the commission's use of Article 90(3) to issue a directive without council consultation was improper, and that the directive itself violated the treaty. In both instances, the Court of Justice vehemently supported the commission's actions and added legitimacy to its approach to redefining the norms in the telecommunications sector. France's first argument was that the telecommunications sector should be grandfathered, because Article 90 could only be applied prospectively, and thus could not be invoked to attack monopolies that were in place before the treaty was established. The Court of Justice dismissed this argument by citing Section (1) of Article 90, which directs member countries to "neither enact nor maintain" rules that permit monopoly holders to engage in anticompetitive behavior to preserve their market power. France went on to argue that under Article 100A of the treaty, the commission was not empowered to take actions to establish the Common Market without the approval of the council and the sanction of the parliament and the Economic and Social Committee. Once more, the Court of Justice supported the commission. In doing so, it drew a distinction between general decision-making power regarding establishment of the Common Market, which it deemed to be subject to Article 100A, and specific enforcement powers in competition matters placed under commission jurisdiction by Article 90. This decision was a victory for the commission, and bestowed upon it new levels of authority. It clearly legitimized its use of competition policy to create and install new Communitywide rules by fiat. But the Court of Justice went further by affirming the principles of a well-known competition case, making it applicable to telecommunications equipment, and leaving the door open to its application to services. In *Procureur du Roi v. Dassonville* [20], it was firmly established that "all trading rules enacted by Member States which are capable of hindering, *directly or indirectly, actually or potentially*, intra-Community trade" were illegal under Article 30 of the treaty. The court found that preserving equipment monopolies would hinder intra-Community trade because no company could possibly keep abreast of all the technological innovations in the telecommunications field, and their continued existence would therefore prevent consumers from obtaining the most modern equipment at competitive prices from companies that manufacture their products outside the borders of the country where they reside. From this conclusion, the court made a leap and declared that the monopolies would therefore have the same effect as a quantitative restriction on trade, and should be illegal under the *Dassonville* principle.

12.3.2 The Services Directive

Even though the commission won the Equipment Directive battle, the vigor with which that directive was contested made it rethink its approach to implementing its competitive agenda. While not ready to give up its newly confirmed authority, it agreed to consult the European Council and the parliament before issuing the final version of its Services Directive [21]. Thus, the interim version of the directive was issued in 1988, and the final version in 1990 [22]. The Services Directive abolished all monopolies except voice and imposed nondiscrimination and transparency obligations on regulators and licensing authorities, and mandated separation of the regulating and licensing authority from the telecommunications operator. On the other hand, it ordered PTTs to bring their prices into alignment with cost in compliance with the ITU's guidelines. The alignment of price with cost was not particularly controversial, as in 1988 the council itself had issued a resolution recommending that the ITU's guidelines on the matter be followed [23]. It also mandated the publication of all interconnection interfaces so that private operators of value-added networks may interconnect with the public switched network. The directive does not apply to services other than voice telephony. Although it permitted the PTTs to reserve their monopoly over voice, this service was given a very restrictive interpretation. Voice telephony was defined as the direct transport and switching of speech in real time between points in the public switched network. Thus, the reservation applied neither to private lines nor to private networks. When a controversy arises as to whether a service is encompassed within this narrow definition, the burden of proof is placed on the PTT. The commission preserved the voice monopoly because it found that dismantling it at that stage could threaten the financial stability of the PTTs. This concern was particularly relevant, since some of those entities were beginning to consider privatization.

As a compromise between the commission and the member states, the Services Directive was tied to a Council Framework Directive on ONP (see below) mandating nondiscriminatory and efficient access to the public switched network by users and providers of telecommunications services [24].

In spite of the compromise and the outcome of the Equipment Directive appeal, Spain, Belgium, and Italy challenged the Services Directive on grounds similar to those alleged for the Equipment Directive [25]. Once more, the Court of Justice upheld the commission's actions and approach. In spite of the compromises of the Services Directive, total liberalization of voice services remained very much on the commission's mind.

In 1991, the commission adopted guidelines to clarify its approach to telecommunications competition based on general European law principles and commission jurisprudence. However, the guidelines also allowed the commission to impose fines of up to 10% of global turnover on companies and governments that do not comply with its competition rules [26]. Not content with that, the commissioner for competition, Sir Leon Brittan began saying publicly in 1992 that he saw no reason to preserve voice monopolies. Sir Leon compared the European situation to the U.S. environment before divestiture and concluded that the European market was inefficient. Sir Leon made those declarations in the course of a visit to Spain, one of the countries that most fiercely resisted the prospect of competition. One month later, the commission began to consider classifying long-distance and international service as services separate from basic local telephony and exploring the possibility of introducing competition at that level. That proposal was withdrawn in April of 1993 because it lacked the support of the member states. France, for instance, complained that competition would force France Telecom to raise the price of local calls (currently still subsidized by high long-distance rates), considered a basic public utility to the detriment of small businesses. Instead, the commission decided to pursue a more gradual approach to liberalization [27].

12.3.3 The Telecommunications Review

The commission then issued a communication to begin the debate about the possibility of opening voice to competition [28]. In the *Telecommunications Review*, the commission identified certain bottlenecks that stood in the way of competition. The first and foremost was tariff structures. The commission found that most operators had done nothing to bring collection rates into alignment with costs. The disparity in cost alignment between member states led to market anomalies such as widely divergent prices on certain routes, depending on the direction in which the call was made. Another important bottleneck was delay in obtaining service and in implementation of new technologies. With its decision-making power firmly established, the commission moved ahead with its plan, this time in consultation with industry participants. Several contributions to the study suggested that, as had been the case in the United States, lost revenues due to price reductions would be amply offset by increased volumes of calls, and lost market share due to competition would be compensated for by growth of the market. The final version of that work appeared in 1993 [29] and recommended liberalizing voice telephony.

Soon afterwards, the council adopted a resolution supporting liberalization of all voice telephony by 1998. A few countries with less developed networks, such as Greece, Ireland, Portugal, and Spain, were allowed to take a derogation until 2003. However, Spain voluntarily renounced the derogation in 1994 [30]. That country has begun to increase its vigilance over competition in the telecommunications sector. In February of 1995, the Spanish Competition Tribunal fined the voice monopoly, Telefónica de España, $1 million for refusing to allow a competitor access to its network.

12.3.4 The Open Network Provision Framework Directive

The ONP Framework Directive (ONP Directive) seeks to help new entrants into the telecommunications marketplace by preventing actual or former monopolies from abusing their position and control of the network to gain or preserve market share to their detriment. It is a set of technical guidelines and procedures defined jointly by the commission, the council, the Senior Officials Group on Telecommunications (SOG-T) [31], and several standardization institutes, including the European Technical Standards Institute (ETSI).

Former monopolies must afford newcomers reasonable access to the public network they control. They may not restrict access to their networks by competitors, must publish the interface standards that make interconnection possible, and make their network available without discrimination and on a tariff basis. Conditions for access to the network must be based on objective criteria. Access can only be denied in case of threat to the security or integrity of the network and for data protection. The ultimate aim of the ONP Directive is to harmonize the networks of all the member states to facilitate the creation of a pan-European network and Telecommunications European Networks (TEN) [32].

The ONP Directive established a set of priorities in accomplishing its goals. The first one was the application of ONP conditions to leased lines and their general availability by 1993, followed by switched data services, ISDN, voice telephony, telex service, and mobile services. In 1993, the council issued a directive concerning leased lines, imposing ONP by June of 1993 [33]. However, the ONP provisions do not apply if the lines originate or terminate outside the EU. In addition, it stops short of mandating dialing parity and number portability in their efforts to create a level playing field for market newcomers.

12.3.5 Infrastructure Liberalization

In December of 1994 and January of 1995, the commission issued a two-part Green Paper proposing to liberalize infrastructure, the last bastion of the telecommunications monopolies. The infrastructure Green Paper is the natural next step after the ONP Directive. The commission wants to liberalize infrastructure by 1998, at the same time competition is introduced in the public switched voice markets.

The Green Paper recognizes natural jurisdiction over licensing of network, services, and infrastructure, as well as over interconnection and interoperability agreements and standards. However, the national licensing procedures and interconnection and interoperability agreements are subject to European competition law and to a Europewide hammorization framework.

The infrastructure paper recognizes the importance of adhering to the World Trade Organization/General Agreement in Trade in Services and of ensuring reciprocal rights for European companies so they can compete in global markets. Directives are expected by the end of 1995.

Meanwhile, the commission also turned its attention to satellite issues.

12.4 SATELLITES

In 1988, one year after the publication of the Green Paper on telecommunications, the commission turned its attention to satellites [34]. Because satellite service is relatively new and there are no historically entrenched monopoly providers, this area has proved less controversial than service and equipment liberalization. The commission's first step was to issue a communication [35] describing the need for a coherent satellite policy. A satellite Green Paper followed in 1990. The satellite paper recommended the total liberalization of the Earth segment. It proposed that systems not connected to the public switched network be regulated only as necessary to avoid harmful interference, protect privacy, and avoid piracy. Uniform type approval procedures and standards should be adopted for transmit-receive terminals. Licenses granted in one country should be recognized in the other member states. The paper also recommended that member states cease restricting access to the space segment of intergovernmental organizations, such as INTELSAT and EUTELSAT. This second provision proved more controversial, and it has only been adopted by the United Kingdom (where all users may have direct access to INTELSAT) and, on a limited basis, in Germany (where users are free to access the system through the signatory of their choice, regardless of its nationality). Another important recommendation was the separation of the regulatory body from the operator in each country in accordance with the Services and ONP Directives. Satellite service providers should be free to market their services directly to users. In supporting the goals of the satellite paper, the council added its own recommendations to accelerate the deployment of separate systems [36].

But the two most important measures adopted by the EU have been mutual recognition of terminal equipment, Earth station equipment, and mutual recognition of licenses. Of these, only the first can be said to be fully accomplished. In 1991, the European Council adopted a directive mandating recognition by all member states of any terminal equipment recognized by the competent authority of any member state [37]. Once equipment is approved by one state, all the others must accept that approval and permit the equipment to be marketed in their jurisdiction and connected to the public switched network. A proposal for the mutual recognition of Earth station equipment followed in 1992. Also in 1992, the commission issued a proposal for a directive concerning mutual recognition of satellite service licenses [38]. Under this proposal, all member states would have to recognize licenses granted by other member states upon application. In the second phase of this proposal, license applications and licensing requirements would be harmonized. The United Kingdom, France, Germany, and the Netherlands already have such a system in place. A single application in one of those countries is all that is currently required [39].

12.5 MOBILE SERVICES

The commission issued a Green Paper on mobile telecommunications in April of 1994, in which the commission recommended eliminating all restrictions on the position of mobile services, permitting providers to build their own networks or use that of other providers, including the PTT's, and unrestricted offering of all services, including voice, on the continued fixed mobile network by 1998. The commission also expressed the view that discretionary licensing of mobile communications operators is illegal. It also warned PTTs against leveraging their power in the landline market into the mobile market. The leverage is enhanced by the intrinsic concentration in this service and the fact that 95% of all calls either originate or terminate in the regular network [40]. Some member states considered the Green Paper a premature pronouncement on infrastructure. The commission has not yet made a recommendation concerning infrastructure liberalization.

With respect to mobile satellites, the EU is worried about lagging behind the United States. All the currently licensed systems will be built by U.S. companies. The commission recognizes the need to adopt a joint European position on this issue and to harmonize the licensing conditions for these services. At the same time, the commission recommends continued consultation with non-EU countries and close monitoring of regulatory proceedings in the United States [41].

12.6 TELECOMMUNICATIONS TODAY

The European telecommunications picture, then, features widespread but increasingly nervous voice monopolies and a few competitive providers of private-line, value-added, and cellular services. The days of voice monopolies are undoubtedly numbered. The old PTTs must privatize and prepare to face competition from companies within and without the EU. Some, including government-owned France Télécom, worry that the impending introduction of competition will be deleterious to their ability to obtain capital in financial markets through public offerings [42]. Others, such as Spain's Telefónica, which has been a largely privately held company since the 1930s, have responded by aggressively expanding internationally [43].

European telephone penetration is lower than in the United States, and market growth is consequently higher [44]. Voice will remain a monopoly until

1998, but other services have been substantially liberalized. The Netherlands and France [45] have allowed other companies to build alternative networks. The United Kingdom undoubtedly has the most liberal regime. It was the first country to permit international resale of private lines interconnected to the public switched network on a reciprocal basis. In late 1994, the United States and the United Kingdom each found that the other country offered equivalent resale opportunities, and gave each other's carriers permission to resell interconnected private lines. In early 1995, the United Kingdom granted a license to AT&T to provide domestic service within the country.

In spite of that process, and with the notable exception of the United Kingdom, there are only a few providers in each liberalized market, such as cellular, and even there the former monopoly remains the dominant firm in most instances. The EU has adopted a digital standard, the Global System Mobile (GSM) standard for cellular communications. While most European cellular markets are duopolies, the uniform adoption of the GSM digital standard and the further removal of national barriers will soon bring more competition.

Europe is witnessing a growth of private networks thanks to the nondiscriminatory interconnection requirements imposed by the ONP Directive. That directive effectively legalized bypassing the public switched network and imposed on the member states the obligation to make available several types of leased lines without restriction on their use, even for voice services. In addition, the high per-minute charges of entrenched European operators make private networks all the more attractive financially.

There is no uniformity of tariffs across Europe, and calling prices are substantially higher than in the United States. A number of studies have concluded that there is considerable pentup demand for more and more advanced communications services repressed by the current tariff structure. This in turn depresses the revenues of European operators. For example, it is estimated that, in spite of lower prices in the United States, the revenue per main line averaged $819 in Europe and $1,200 in the United States [46]. The high European collection rates have also fueled the growth of alternatives to the national public switched networks in order to avoid those high prices, such as private networks, virtual private networks, and call-back services. Virtual private networks are becoming increasingly popular in the United States as the price of calls is reduced and technology advances. They consist of software-defined networks that give a client a dedicated line upon demand. Routing of the call of information being transmitted is not uniform. Rather, the message will be transmitted through any facilities available at the time. Because it avoids true dedicated lines, this system is more economical to operate. The ONP Directive forced telecommunications operators to make their facilities available to competitors seeking to offer virtual private networks. Because they involve a closed group of users, these communications are outside the exclusive competence of PTTs. In Europe, then, virtual private network operators route the calls through the networks of the operators offering the lowest prices at substantial discounts over regular rates [47]. Call-back services consist of software that can trace a call. A European caller dials a number in the United States and hangs up after three rings. Because there has been no answer, no charge is made. On the U.S. side, a computer is able to identify the station used by the caller, and calls him or her back immediately with a U.S. dial tone. The caller can then place calls to any number and be billed at U.S. rates. Because telephone tariffs vary so widely across Europe, a number of companies use call routing to save money on the price of calls. Since intra-European generally cost less than calls outside the EU, a savvy company can save substantial amounts of money by routing its traffic by private line to the United Kingdom (where the price is lowest) and then forwarding it outward from there.

In anticipation of competition from U.S. firms, a consortium composed of France Télécom, British Telecom, Deutsche Telekom, Telefónica, and ASST/STET installed the first General European Network (GEN) in the Spring of 1993. This GEN will provide private lines between Paris, London, Frankfurt, Madrid, and Rome.

12.7 THE LIMITS OF INTEGRATION: THE ANTITRUST GUIDELINES

One of the side effects of the ambitious integration and development plans of the EU has been necessary cooperation between telecommunications enterprises in their pursuit of pan-European networks and services. The GEN referred to above is a good example. The EU has recognized the need to watch over these schemes to prevent cooperation between enterprises from resulting in new monopolies or cartellike arrangements that would compromise competition and the very goal of the 1987 Green Paper: to bring a wider range of better and cheaper telecommunications services to European users.

Cooperation agreements between telecommunications operators are essential to accomplish the commission's goals of interoperability, interconnectivity, one-stop shopping, harmonization, and integration. In early 1994, the commission published nonenforceable guidelines concerning cooperation between operators [48]. In these guidelines, the commission offers guidance to operators as to "which forms of cooperation amount to undesirable collusion," so that they may plan and finance their networks and services with greater regulatory certainty. The guidelines now apply to all EFTA countries, and not just to members of the EU. They apply regardless of whether the operator is owned by the government or privately held.

The commission reviews all substantial mergers and cooperation agreements [49]. In order to assess the effects of an agreement on competition, the relevant product/service market and geographical market must be defined first. Because of the fluid nature of telecommunications products and services, market definition in this area is a variable concept, and the specific market will, in the end, be defined on a case-by-case basis. However, the commission defines the parameters within which the market will be determined to exist as "the totality of the products which, with respect to their characteristics, are particularly suitable for satisfying constant needs and are only to a limited extent interchangeable with other products in terms of price, usage and consumer preference." The determination will take into account "the structure of supply and demand" and competitive conditions in the market [50]. At the very least, the commission found terrestrial networks, voice communications, data communications, satellites, public switches, private switches, transmission systems, mobile telephones, telephone sets, and modems to be separate markets. However, the guidelines recognize that those markets can be divided into further submarkets.

A geographical market is an area where undertakings compete with each other and the objective conditions of competition are similar for all traders [51]. The commission expects to extend the definition of geographical market to encompass the entire EU and the EFTA.

12.7.1 Horizontal Restrictions

The economic integration of the EU has given rise to a new telecommunications phenomenon known as *hub competition*. Because telecommunications is now considered the fourth factor of production [52], the communications options available to an enterprise and their price are increasingly becoming a factor in its decision of where to locate its European headquarters and offices [53]. Certain agreements between operators, can then have the effect of restricting hub competition. The most obvious and egregious examples are price agreements. Hub competition relies heavily on pricing differentials between operators. The guidelines place a very high burden on any operator that proposes to enter into a price agreement to demonstrate why it should not be struck down. Since discounting is the primary form of competition in most instances, any restriction on an operator's ability to offer discounts will be highly suspicious. On the other hand, the guidelines encourage the adoption of common tariff structures or principles as an economic benefit whose anticompetitive effects are outweighed by its economic benefits [54], such as harmonization and drive toward cost-based tariffing.

Agreements as to the choice of routing for services are suspicious because of their clear potential effect on tariffs (as to the end user). They are also an effective way to divert traffic from competitors.

While the guidelines generally favor common technical and quality standards, they recognize their potential anticompetitive effects: "hindering, freezing a particular stage of technical development, blocking the network access of some users/service providers" [55]. In examining such agreements, the commission will conduct a complicated analysis to balance the economic benefits of the proposal against its anticompetitive effects. Standards promulgated under the

ONP Directive enjoy a presumption of legality that can only be rebutted by a demonstration that the agreement contains restrictions that are not indispensable to the standardization. There is an additional presumption of legality when users and manufacturers have been involved in developing the consumer standards.

Agreements on the exchange of proprietary information, including business strategy, are prohibited.

12.7.2 One-Stop Shopping Agreements

The purpose of these agreements is to offer to customers the benefits of an entire seamless package of service. It involves the provision of facilities, services, and management. One of the goals of the 1987 Green Paper was to promote the development of these arrangements. These agreements will constitute a restriction on competition if they result in coordination on prices charged to end users; provide for joint specifications of products, quotas, or delivery; or they contemplate joint purchase of software or hardware. Naturally, any restriction on an entry by a third party (such as refusal to provide facilities, usage restrictions, less favorable terms, or cross-subsidization) are illegal.

Agreements concerning joint research and development will be looked upon favorably as long as they do not contemplate joint distribution.

12.7.3 Abuse of a Dominant Position

Most former telecommunications monopolies enjoy a dominant position in each national market with respect to the position of network services. Market power is defined in terms of control of the market as evidenced by exclusive concessions (e.g., a voice monopoly) or by sheer market share. The guidelines are concerned with leveraging a dominant position in one market to expand into or acquire control of another. As above, the above can take the form of usage restrictions or prohibitions, or with acceptance of unreasonable conditions not related to usage. Refusals to supply and to deal have been considered abuses by the European Commission and the Court of Justice, especially in an essential facilities context [56]. Imposition of extra charges or conditions is also unduly discriminatory.

12.7.4 Mergers

The guidelines recognize the beneficial effect of mergers to achieve economics of scale and scope and to permit financing advanced research and development. Thus, merger candidates must demonstrate that the economic benefits of a merger outweigh its anticompetitive effects. The guidelines caution against strengthening dominant positions by way of mergers. In this context, it will not be enough to argue that the merger would make the European participant more competitive in world markets.

For example, in 1994, the commission cleared the acquisition of 20% of the U.S. telecommunications company MCI by British Telecom. The commission found that although British Telecom would become the largest single shareholder of the company and would obtain certain exclusive distribution rights, U.S. antitrust and corporate law would prevent abuse of its leverage.

Notes

[1] Helmut Ricke, "Germany's Telekom: A New Way of Doing Business in a Liberalized Market," *Telecomm. J.*, October 1991, p. 714.
[2] Mail, telegraph, and telephone.
[3] Europeans are particularly fond of a World War II anecdote, which is not entirely accurate. According to it, the United States became aware of the possibility of an impending Japanese attack on Pearl Harbor, and the High Command immediately sent a wire to Admiral Kimmel of the Pacific Fleet. Having no reliable military network, the High Command sent the message through Western Union. The private company misplaced the wire, and the message was not delivered until the attack had already begun.
[4] Its members were Belgium, France, Germany, Italy, Luxembourg, and the Netherlands.
[5] It was later approved in a second referendum.
[6] Merger Treaty, Article 11.
[7] This provision was adopted at the insistence of France in 1966 and is known as the Luxembourg Accords.
[8] Commission Decision of December 12, 1982, O.J. (L 360) 36.
[9] Case 41/83, *Italy v. Commission*, 1985 E.C.R. 873; EEC Treaty, Arts. 86, 222.
[10] J. E. Darnton and D. A. Wuersch, "The European Commission's Progress Toward a New Approach for Competition in Telecommunications," *Int'l Law*, Vol. 26, 1992, p. 111.
[11] COM(87) 290 Final (1987).
[12] Commission of the European Communities, *Towards a Dynamic European Economy, Summary Report Concerning the Green Paper on the Development of the Common Market for Telecommunications Services and Equipment.*
[13] Green Paper, Figure 3 at B.
[14] Green Paper, p. 3.
[15] Committee of the European Communities, *Towards a Competitive Community-Wide Telecommunications Market in 1992: Implementing the Green Paper on the Development of the Common Market for Telecommunications Services and Equipment*, COM(88) 48, p. 13.
[16] Council Resolution of June 30, 1988 on the Development of the Common Market for Telecommunications Services and Equipment up to 1992.
[17] Commission Directive 88/301 on Competition in the Markets in Telecommunications Terminal Equipment, 1988 O.J. (L131)73 (hereafter called Equipment Directive).
[18] Commission Directive 90/388 on Competition in the Markets for Telecommunications Services, 1990 O.J. (L192) 10 (hereafter called Services Directive).
[19] Case C-202/88, *France v. Commission*, 5 C.M.L.R. 552 n. 14 (emphasis added).
[20] Case 8/74, 1974 E.C.R. 837, 2 C.M.L.R. 436 (1974) (emphasis added).
[21] See A. Winter et al., *Europe Without Frontiers: A Lawyer's Guide*, 1990, pp. 238–239.
[22] June 28, 1990.
[23] Council Resolution of June 30, 1988 on the Development of the Common Market for Telecommunications Services and Equipment, 1988 O.J. (C 257) 1.
[24] Council Directive 90/387 on the Establishment of the Internal Market for Telecommunications Services, 1990 O.J. (L192) 1,3.

[25] *Spain, Belgium and Italy v. Commission,* 271/90, 281/90 and 289/90 (joined cases).
[26] Guidelines on the Application of EEC Competition Rules in the Telecommunications Sector, 1991 O.J. (C233) 2.
[27] Commissioner Karel Van Miert announced this decision in April 1993. *Telecommunications Reports*, April 19, 1993, p. 10.
[28] 1992 Review of the Situation in the Telecommunications Sector, SEC(92) 1048.
[29] Consultation on the Review of the Situation in the Telecommunications Services Sector, COM(93) 159.
[30] Thus, Telefónica's monopoly will end January 1, 1998. *Communications Daily*, October 1, 1994, p. 7.
[31] This is a group composed of the telecommunications authorities of the member countries. In providing its input, the group consults telecommunications users, consumers, manufacturers, and providers of services.
[32] The Maastricht Treaty mandated the development of Europewide TENs.
[33] Council Directive on the Application of Open Network Provision to Lease Lines, 92/44/EEC, O.J. L. 165/27, June 19, 1992.
[34] For a comprehensive treatment of European satellite issues, see S. White, S. Bate, and T. Johnson, *Satellite Communications in Europe, Law and Regulation*, London: Longman, 1994.
[35] Communication on Space: The Community and Space: A Coherent Approach, COM(88) 417.
[36] This is a term referring to systems other than INTELSAT.
[37] Approximation of the Laws of the Member States Concerning Telecommunications Terminal Equipment, Including the Mutual Recognition of Their Conformity, Council Directive 91/263/EEC, O.J. L128/1, 5/23/91.
[38] Proposal for a Council Directive on the Mutual Recognition of Licenses and Other National Authorizations for the Provision of Satellite Network Services and/or Satellite Communications Services Extending the Scope of Directive EC, Commission of the European Communities: DGXIII/231/92 Rev 3EN.
[39] *Communications Daily*, April 28, 1993.
[40] Green Paper on a Common Approach in the Field of Mobile and Personal Communications in the EU, April 27, 1994.
[41] S. M. Taylor Article, Telecommunications Monopolies, 15 E.C.L.R. 322, 235, November-December 1994.
[42] *L'Express*, February 9, 1995, p. 7.
[43] To this day, Telefónica has holdings in Argentina, Chile, Peru, and the United States (Puerto Rico).
[44] U.S. Congress, Office of Technology Assessment, *U.S. Telecommunications Services in European Markets,* OTA-TCT-548, Washington, D.C.: U.S. Government Printing Office, August 1993.
[45] Compagnie Générale des Eaux (CGE) and Société Francaise de Radiotéléphone (SFR). G. C. Staple, ed., *Telegeography 1994*, p. 38.
[46] Commission of the European Communities, *Towards Cost Orientation and the Adjustment of Pricing Structures—Telecommunications Tariffs in the Community*, Brussels, July 15, 1992, cited in U.S. Congress, Office of Technology Assessment, *U.S. Telecommunications Services in European Markets*, OTA-TCT-548, Washington, D.C.: U.S. Government Printing Office, August 1993.
[47] G. Finnie, "Competition in Europe," in *Telegeography 1994*, G. C. Staple, ed., p. 42.
[48] C.M.L.R Antitrust Reports, February/March 1994, p. 363 et seq.
[49] Under a process known as *notification*. The commission must be appraised of a proposed agreement or transaction. Before issuing its ruling, the commission solicits public comment.
[50] Thus incorporating European jurisprudence. See Case 322/81, *Michelin v. E.C. Commission* (1993) E.C.R. 3461, (1985) C.M.L.R. 282, para. 37.

[51] See Case 27/76, *United Banks v. E.C. Commission* (1978) E.C.R. 207, (1978) I.C.M.L.R. 729, para. 44 and Case 247/86, *Alsatel-Norasam* (1988) E.C.R. 5987 (1990) 4 C.M.L.R. 434.
[52] Together with land, labor, and capital.
[53] See V. Pastor, "The International Accounting Rates Problem: The European Commission Steps In," *FCLJ*, Vol. 45, 1993, p. 313.
[54] Under Article 53(d) of the treaty.
[55] Ibid., p. 382.
[56] Cases 6-7/73, *Commercial Solvents v. E.C. Commission* (1974) E.C.R. 223 (1974) I.C.M.L.R. 309, *United Brands v. E.C. Commission.*

13 Russia, the Former Soviet Republics, and Eastern Europe

One of the central events of our time is the fall of the USSR and its East European satellite regimes. Among other legacies, these governments left in their wake an empire of outdated, state-run enterprises, including some of the most primitive telephone systems in the industrialized world. In order to understand the task faced by these countries as they seek to modernize and reform their telecommunications sectors, we should understand the peculiar features of the system they inherited from decades of collectivist rule.

One such feature was the mixture of indifference and hostility with which Marxist regimes viewed communications technology. As a matter of economic planning, communications and other services were a distraction from the serious work of heavy industrial production [1]. As a matter of state security, telephones were viewed with suspicion and low penetration rates, far from causing concern, were desirable [2].

Another feature of the old system was the restrictive, barter-style set of economic relationships between the Soviet Union and its satellites, formalized in the Comecon system that persisted until 1991. Under Comecon, the development of trade in hard currencies was retarded and telecommunications equipment manufactured in Eastern Europe went primarily to the USSR in exchange for oil and other Soviet commodities. These restrictions, combined with the refusal of the West to sell advanced communications technologies to the Warsaw Pact countries [3], made it all the more difficult to upgrade those countries' networks.

Finally, the Eastern Bloc countries had the problems common to all countries that provide telephone service through state-owned monopolies, including inefficient management and the practice of returning revenues to the public treasury rather than investing them in network improvements and expansion. These problems were exacerbated by the absence, in Communist countries, of private enterprises against which the PTTs' performance could be measured and from which sound business practices might be borrowed.

The results of this heritage are a discouraging prospect. From the Baltic states to the Balkans and from Warsaw to Vladivostok, telephone penetration rates are low, waiting lists for service are long, and wireline transmissions are routed by manual and electromechanical switches that would be museum pieces in the West. The situation is so difficult, in fact, that the ITU estimates a cost of $50 billion to $60 billion simply to bring penetration rates up to the level of Ireland, which has one of the lowest telephone penetration rates in Western Europe—and that number assumes only the addition of new lines, without replacements or upgrades of existing facilities [4].

As the balance of this chapter shows, the East European telephone administrations generally have announced their intention to privatize their PTTs, bring their tariffs into line with costs, and open their markets to new investment and competition. However, efforts to implement these plans have met with a number of economic and political obstacles.

Most of the economic obstacles stem from the low productivity and foreign exchange reserves of the East European economies, which are unequal to the size of the task. To compensate for these shortcomings, the countries in the region are working to attract foreign capital. However, private investment so far has gone disproportionately to a few countries and predominantly for wireless and other business-oriented services [5], leaving modernization of the wireline networks to be financed primarily from internal sources.

Political obstacles include the actual or perceived instability of some of the regimes in the region [6], resistance to foreign investment [7], and the desire to protect faltering internal industries and the associated employment [8]. Fortunately, the desire of East European nations to become full participants in the EU gives the proponents of liberalization some much-needed leverage in these internal political debates [9].

The discussion that follows surveys the efforts at liberalization made in several nations of Eastern Europe and the former USSR.

13.1 RUSSIA

Even after the collapse of its empire, Russia remains a vast and powerful continental nation. With its enormous oil reserves, its huge (if inefficient) industrial base and its well-educated population, Russia can absorb a great deal of political and economic punishment without entirely squandering its claim to great-power status.

During the decades of totalitarian rule by the troika of the Communist Party, the Red Army, and the KGB, the Soviet Union developed three telephone systems. One system, available only to high-level civilian and military officials, was secure and could be accessed only with special keys; another system was used only by the Communist Party; and the third system, available to the public, was characterized by low penetration, lack of reliability, and long waiting lists for service [10]. Provision of public telephone service was the responsibility of the Soviet Ministry of Posts and Telecommunications and a number of local telephone administrations.

Beginning with the era of peristroika and continuing to the present, Russia has moved gradually to restructure and privatize its telecommunications sector. The process began in 1990, when the Council of Ministers voted to turn the sector over to a joint stock company called Sovtelecom (later called Intertelecom).

As the Soviet Union dissolved, with many of the republics declaring their independence and taking their portion of Intertelecom's assets with them, the Russian Federation decided to privatize the Russian telecommunications sector. In preparation for this effort, the government separated the sector into several regional carriers and one international and long-distance carrier. The latter company was called Rostelecom.

Russia is now in the process of privatizing both Rostelecom and Svyainvest, the holding company that owns 85 of the local and regional companies. Rostelecom will be held 38% by the Russian government, 40% by the company's employees, and 22% by the public. As of October 1995, the government was planning an initial offering of 25% of Svyainvest to a group of foreign telecommunications companies.

International and long-distance competition have begun in Russia, with Svyainvest and other companies already obtaining licenses to compete with Rostelecom. The Ministry of Posts and Telecommunications regulates the entire industry and sets long-distance tariffs, but local ministries have the primary responsibility of regulating local rates. There are, as yet, no detailed rules to govern pricing and interconnection.

Most of the new capital coming into the Russian telecommunications sector, however, is not going to the improvement and expansion of basic service, but is dedicated to the construction of overlay, cellular, international, and other projects that are chiefly of value to large business customers [11]. Those projects are owned and managed by a variety of public entities, private companies, and public-private joint ventures. (The few modernization projects that focus on improvements to existing local networks, rather than construction of overlays, tend to be in Moscow and other major business centers [12]).

While Russia has moved rapidly to award cellular licenses, augment its international gateways, and build overlay networks connecting its major business centers, it has made almost no progress in modernizing its basic public network. (This problem will not be solved, of course, until Russia's economy achieves the productivity required to generate sufficient capital to support this enormous effort.) Russia also lacks a coherent system of regulation for the many entities involved in providing telecommunications services—a deficiency that increases the risk of companies operating in Russia and may act as a curb on foreign investment.

Finally, Russia's continued ability to attract capital and expertise and to integrate its telecommunications system with that of the rest of the developed world will depend in part on its political stability and the direction of its foreign policy. Already, the West's relationship with post–Cold War Russia has come under some strain over the war in Chechnya and Russian sympathy for Serb ambitions in the former Yugoslavia. Frictions of this kind, or a resurgence of militarism in the Russian government, could revive the CoCom restrictions and restore some measure of the economic isolation that retarded economic development in the USSR.

13.2 FORMER SOVIET REPUBLICS

When the Bolsheviks seized power in 1917, they inherited the empire of the Czars—a huge collection of territories encompassing a wide range of religions, languages, and cultures. After a period of civil war and reconsolidation, the Communists named this empire the Union of Soviet Socialist Republics, added to it some additional conquests, and subjected the republics to a campaign of economic, cultural, and religious suppression that was long, brutal, and ultimately unsuccessful.

When the Gorbachev regime finally relaxed Moscow's grip on the republics, the immediate result was a loose confederation called the Commonwealth of Independent States. The long-term viability of this confederation is uncertain, although in certain areas—including telecommunications—some coordinated planning among the republics has proved to be a practical necessity [13]. Otherwise, the former Soviet republics have evolved rapidly into fully independent, sovereign nations with their own approaches to infrastructure development and other problems [14]. The following reviews the efforts of some of these nations to expand and liberalize the telecommunications sectors they inherited from the Communist regime.

13.2.1 The Ukraine

After Russia, the Ukraine is the most populous of the former Soviet republics, with a population of over 50,000,000 and a territory the size of France. Like the other republics, the independent Ukraine inherited the primitive, state-owned telephone system of the USSR.

The Ukraine now has its own Ministry of Communications, which so far has concentrated on finding joint venture partners for long-distance, international, and cellular services. The ministry has not organized a specialized regulatory agency or promulgated pricing, access, or other rules to govern the evolving telecommunications environment.

The Ukraine is attempting, through the terms of its concession with joint venturers operating in the country, to serve the needs of basic subscribers as well as large business customers. Specifically, the Utel joint venture among AT&T, PTT Telekom of the Netherlands, Deutsche Telekom, and the Ukraine State Committee of Communications is required to boost the number of local telephones in the country from 7,000,000 to 13,000,000, as well as building new international and long-distance facilities.

The Ukraine has had some problems with inconsistent regulation. The Utel partnership, for example, was originally assured of a 15-year exclusive franchise, but within two years a second license had been issued, subjecting Utel to competition. Problems of this kind, which do not reassure investors, demonstrate again the wisdom of having a basic regulatory framework intact at the beginning of the telecommunications reform process.

13.2.2 Armenia

The present Armenian Republic occupies part of the territory of ancient Armenia, which once included areas now lying within Turkey and Iran [15]. It has a population of about 3,300,000 people living in a region of the South Caucasus bordered by Georgia to the north, Azerbaijan to the east, Turkey to the west, and Iran and Turkey to the south.

The Armenian Republic's efforts at infrastructure development have been delayed by more immediate concerns, including territorial disputes with neighboring republics. Armenia does have a Ministry of Communications with full authority to regulate the sector and a policy that favors the licensing of new carriers (including foreign carriers) to provide local, long-distance, and international service. At this writing, the ministry has been corporatized (i.e., regulatory and operational functions have been separated), but neither privatization nor the development of a regulatory scheme for the sector has moved beyond the planning stage.

13.2.3 Kazakhstan

Kazakhstan, the second largest former Soviet republic after Russia, is a nation of about 17,000,000 people bounded by China on the east and the Caspian Sea on the west. With large oil and gas reserves and an ethnic Russian population of around 42%, Kazakhstan will continue to receive close scrutiny from the makers of Russian foreign policy.

Kazakhstan's Ministry of Posts and Telecommunications was corporatized in 1993, and the operating section of the ministry was privatized in that same year. The government's long-range goal is to introduce competition into all phases of the telecommunications sector, but progress toward that goal is rudimentary so far.

13.2.4 Georgia

Georgia, a republic of about 6,000,000 people on the northern borders of Turkey and Armenia, was briefly independent after the Russian Revolution and incorporated into the USSR by the Red Army in 1922. Since the breakup of the USSR, Georgia has been plagued by internal warfare and civil disorder, which has retarded the task of infrastructure development [16].

Georgia plans to corporatize and privatize its Ministry of Telecommunications and to license competitive providers of local, long-distance, and international services. In preparation for privatization, the ministry has raised service rates to approximate the cost of service and finance expansion and modernization of the network.

13.2.5 Belarus

Belarus, the successor state to the Belorussian Soviet Socialist Republic, is a nation of about 11,000,000 people bordering on Russia, Latvia, Lithuania, and the Ukraine [17].

After its independence from the USSR, Belarus continued to operate all telecommunications through the Ministry of Posts, Communications, and Informatics, which exercises both regulatory and operational functions. The terms of a loan from the European Bank for Reconstruction and Development, however, require corporatization of the ministry. At present, the government has no other definite plans to liberalize telecommunications.

13.2.6 The Baltic States

Estonia, Latvia, and Lithuania were late additions to the Soviet empire. They were occupied by the Red Army in 1940, invaded by Nazi Germany in 1941, and reconquered by the USSR in 1944. Their independence was restored in 1991.

Since 1991, the three Baltic states have moved well ahead of the other ex-republics in disentangling their telecommunications sectors from that of the Soviet Union and privatizing and modernizing their networks. The following sections summarize the progress made by each of these states.

Estonia

Estonia passed a new communications law almost as soon as it declared its independence in 1991. The new law separated the post office from the state-owned telephone monopoly (now called Estelcom) and created a new ministry to regulate telecommunications. Estelcom was partially privatized in 1993, with minority investment from Telia AB Sweden and Telecom Finland. With this new investment available, the government plans to bring telephone penetration in line with Western European standards by the year 2000.

The present law gives Estelcom an exclusive concession for basic service until the year 2000 and an opportunity to seek an extension of that concession for another 14 years. Competition is permitted for mobile, paging, and data services.

Latvia

Latvia, like Estonia, separated its post office from its telephone company in 1991, creating a state-owned monopoly telecommunications provider called Lattelkom. Regulatory supervision of the industry was retained by the communications department of the Ministry of Transport.

At this writing, Latvia has not privatized Lattelkom or passed a comprehensive communications law. The government has, however, permitted the construction of private networks and competition in value-added services.

Lithuania

Lithuania passed a new telecommunications law in 1991, creating a Ministry of Communications and Informatics and providing for competition in the provision of local services [18]. In 1992, the government created Lithuanian Telecom (LT), a state-owned enterprise separate from the regulatory function of the ministry, and divided its operations among 10 geographic and 11 functional subsidiaries.

While the government intends to privatize LT, it does not plan to do so in the near future and at this writing no private investors own any part of the company. In the meantime, LT continues to enjoy a monopoly of international and long-distance service. Local, wireless, and value-added services are open to competition.

13.3 HUNGARY

Although Hungary's telephone system is primitive and its penetration rate (17%) is low, its prospects for improvement are better than those of most of its neighbors. Hungary's advantages are its pro-business government and healthy investment climate, which have made it by far the most popular destination for Western capital seeking opportunities in Eastern Europe, and the careful program of privatization and regulation it has followed.

The process of making the Hungarian telecommunications structure more efficient dates from 1987—before the collapse of Communism—when Hungary concluded a loan agreement with the World Bank for expansion of its telephone system. The loan agreement required Hungary's PTT to maintain auditable records of its telecommunications operations separate from its postal functions.

Compliance with World Bank requirements also led to other reforms, such as bringing tariffed rates in line with costs and eliminating the subsidies that flowed from telephone service to postal service.

The Communist Party's monopoly on political power in Hungary ended in 1989, and in that same year the Hungarian PTT's regulatory functions were separated from telephone operations and transferred to the Ministry of Transport, Communication, and Construction [19]. The PTT was renamed the Hungarian Telecommunications Company (HTC), and new legislation permitted minority private ownership of the PTT. Provision of wireless services was liberalized in 1990.

Following the fall of Communism and before the new regime had adopted a legal structure for liberalized telecommunications, HTC took substantial steps of its own to attract capital and expand the network. Notably, HTC took out new loans from the World Bank and other organizations, issued bonds, and created quasiprivate subsidiaries to build local networks. By 1991, these efforts had reduced waiting times for service from 12 years to 5 years and had increased penetration from 8% to 11% [20].

In late 1992, the parliament passed a new telecommunications act for Hungary. The new act creates 56 local service areas, in each of which local service will be provided by a company operating under government concession [21]. (All companies operating under concessions must separate monopoly services from competitive operations to prevent cross-subsidization.) Until 2002, all voice telephone service will be offered by long-distance, international, and several local monopolies. All other telecommunications services are open to competition, subject to government licensing requirements. Tariffed rates are subject to price cap regulation, and a universal service fund will subsidize the provision of service to high-cost areas.

The new telecommunications regime in Hungary continues to be dominated by HTC, which was partially privatized—through sale of a 30% interest to DBK Telekom and Ameritech—in 1993. In late 1994, the government announced that it would sell additional shares of HTC in 1996, but that the government would retain at least a 25% interest.

The government's intention is to preserve monopolies for basic service until its universal service goals are met—specifically, an increase from the present 17% penetration to a 25% or 30% penetration. Given the state of the infrastructure, this goal may not be achieved by the present deadline of 2002 [22].

13.4 POLAND

Poland is a nation of about 38,000,000 people living in a region bordered by Russia and the Baltic Sea on the north, Germany on the west, the Czech and Slovak republics to the south, and Russia to the east [23]. Poland became a satellite nation of the USSR in 1947 and regained its independence during the revolutionary events of 1989 and 1990.

Poland inherited from the Communist era an analog telecommunications network built largely in the 1940s and a penetration rate of slightly better than one telephone for every 10 people. Bringing this network up to Western European standards is estimated to require an investment of $15 to $20 billion.

Poland began the task of liberalization in 1991, when it passed a new telecommunications law establishing an exceptionally open regulatory framework. Under the new law, local competition is permitted and may be provided by foreign-owned carriers. (Long-distance companies must be 51% Polish-owned, and international providers of switched services must be 100% Polish-owned.) The dominant Polish telecommunications carrier, Telekomunikacja Polska (TPSA) is still 100% government-owned, but will be privatized as early as 1996.

Under the liberalized regime for the provision of local service, a number of competitive local service companies have emerged and have formed an Association of Telecommunications Network Operators to act in matters of joint concern. In the meantime, TPSA is using funds from a European bank for a reconstruction and development loan to create a modern overlay network serving Warsaw and other areas. The government hopes to replace aging digital facilities, install new digital capacity, and achieve West European penetration levels by 2005.

13.5 SLOVAKIA AND THE CZECH REPUBLIC

Slovakia and the Czech Republic are the result of the amicable division of Czechoslovakia in early 1993. The combined population of the two nations is about 16,000,000 people, most of whom live in the Czech Republic.

Both republics, of course, inherited their telecommunications sectors from the Communist regime that ruled Czechoslovakia between 1948 and 1989. At the time the nation declared its independence, its telephone system had a penetration rate of 13.6% and 372,000 people on the telephone service waiting list.

Reform of the telecommunications sector began after the end of Communist rule and before the dissolution of Czechoslovakia. Specifically, in 1992 a new federal telecommunications law assigned regulatory authority to the Federal Ministry of Posts and Telecommunications and assigned operating authority to two carriers—SPT-Braha and SPT-Bratislava. This federal arrangement, however, did not survive the 1993 divorce. By late 1992, the new ministry had been replaced by separate ministries in each of the new republics. Each republic then adopted a separate, temporary telecommunications act. The Czech Republic's PTT (SPT-Braha) was renamed Czech Telecom, and Slovakia's PTT (SPT-Bratislava) was renamed Slovak Telecom.

The Czech Republic began the process of privatizing its industry in 1995, when it sold a 27% share of Czech Telecom to a joint venture of Swiss and Dutch PTTs.

13.6 BULGARIA

Bulgaria is a nation of 9,000,000 people bordered on the south by Turkey and Greece, on the west by the former Yugoslavia, on the north by Romania, and on the east by the Black Sea. It became a Soviet satellite nation in 1946 and achieved its independence in the years 1989 to 1991.

Bulgaria entered upon its independence with one of the highest telephone penetration rates (25%) in Eastern Europe. With the assistance of the World Bank, Bulgaria is building a digital overlay network and expects to digitize its core network by 2005.

The present regulatory and operational framework for Bulgaria was established in 1991, when the government created a Committee of Posts, Telecommunications and Informatics and a state-owned monopoly service provider called Bulgarian Posts and Telecommunications. Bulgaria intends to privatize this carrier, but at this writing it remains 100% state-owned and enjoys a monopoly of all but value-added services.

13.7 THE FORMER EAST GERMANY

Since the reunification of Germany, the central government has been faced with the task of integrating the economy and infrastructure of the former German Democratic Republic (GDR)—a Soviet satellite of some 17,000,000 people—into those of one of the most advanced industrial nations in the world. The task is particularly acute in telecommunications: at the time of reunification, East Germany had a telephone penetration rate of only 11% and a waiting list nearly as large as the number of people who had telephones. Most residential subscribers had to rely on party lines, and public mobile and value-added services did not exist.

Germany has moved rapidly to bring the GDR's telecommunications system within the West German regulatory structure. In 1990, the East German regulatory body was absorbed in the West German Ministry of Posts and Telecommunications, and the East German telephone company became part of the West German monopoly, DBP Telekom. The central government also announced an ambitious program, called Telekom 2000, to modernize the East German sector and integrate it fully into the DBP Telekom network.

Merging the two systems is not a simple process. The GDR's Ministry of Posts and Telecommunications not only provided and regulated telephone service, it also distributed all newspapers and journals in the GDR, ran a broadcasting empire, and manufactured telecommunications equipment for domestic use and export. All of these activities had to be disaggregated before integration with the West German network could be attempted. Efforts in this direction began even before unification, when the last of the East German governments worked cooperatively with West Germany to begin rationalizing the ministry and PTT of the GDR [24].

Under the Telekom 2000 program, the East German sector has substantially been integrated with that of the West. This means, among other things, that DBP Telekom provides all services—except nonvoice satellite and mobile services—on a monopoly basis [25]. It also means that tariffs for services in the East and West have been brought into line with one another with sometimes unfortunate consequences [26].

Notes

[1] Soviet-style central planning was a process in which favored programs enjoyed unlimited funding while other sectors of the economy languished at third-world levels. The favored programs changed with political fashion and included Lenin's hydroelectric projects, Stalin's steel mills, Khrushchev's space and ICBM programs, and Brezhnev's deep-water navy. Consumer goods and services were perennial also-rans.

[2] Telephone directories were classified documents in Warsaw Pact countries, and domestic long-distance calling—much less international service—was prohibited or highly restricted.

[3] The West's Cold War trade restrictions were the concern of the Coordinating Committee on Multilateral Export Control (CoCom), which classified digital transmission equipment, optical fiber, and other advanced communication technologies as strategic exports with military applications. The CoCom rules still have some potential to frustrate exports to Russia, but have become largely irrelevant elsewhere in Eastern Europe.

[4] See "New Study Says Eastern Europe, ex-USSR Need to Spend $94 Billion to Upgrade Phones," *International Trade Reporter*, Vol. 9, No. 41, October 14, 1992, pp. 1758–1759.

[5] At one point, half of all the foreign capital reaching Eastern Europe was going to Hungary. U.S. Congress, Office of Technology Assessment, *U.S. Telecommunications Services in European Markets*, 1993, p. 122.

[6] Plans to modernize Yugoslavia's telecommunications sector, for example, were disrupted by the collapse of that country and the descent into warfare of some of the resulting republics.

[7] In Poland, for example, plans to license a foreign-owned cellular telephone system were interrupted when the government withdrew authority for 100% foreign ownership of the system.

[8] As we noted earlier, the CoCom system gave East European countries a ready market for their telecommunications equipment, even when that equipment would not have been competitive on world markets. The collapse of CoCom creates pressure to guarantee access for those industries to domestic markets, even at the price of retarding modernization.

[9] See *U.S. Telecommunications Services in European Markets* [5], p. 126. The desire to achieve rapid parity with the EU countries has also caused some administrations to give priority to cellular systems and overlay networks (based on microwave and optical fiber) that connect major cities and give business customers access to international services. These efforts, too, carry some political risk, since they may be perceived as an abandonment of the interests of ordinary customers.

[10] As this industry structure shows, the notion of overlay networks for favored users is not a new one in Russia.

[11] Potentially the largest of the overlay projects is the 50,000-km fiber-optic network designed to connect 50 digital switches across Russia. This project, which is a joint venture involving two Russian companies, U S West International, France Télékom, and Deutsche Bundespost Telekom, will double the number of telephone lines in Russia. "Partners Give Go-Ahead to Russian 50/50 Project," *Telecommunications Reports*, October 10, 1994, p. 30.

[12] So, for example, AT&T was awarded a contract by Moscow Local Telephone Network and Rostelecom, Russia's principal long-distance company, to expand the Moscow telephone system with digital facilities. "AT&T Gets Moscow Network Modernization Contract," *Telecommunications Reports*, October 3, 1994, p. 22.

[13] Most of the independent states are members of Regional Commonwealth for Communications. This group coordinates telecommunications services among the former Soviet republics, but does not regulate the internal telecommunications of any of its members.

[14] The continuing independence of the republics may be said to be at the sufferance of the Russian government, which continues to regard the republics as within its legitimate sphere of influence. Russia has involved itself in territorial disputes among republics and has used military force to prevent at least one region (Chechnya) from achieving independence. Activities of this kind, along with the tendency of Russia to champion the interests of Russian minorities living in the independent states, could reduce the independence of some of the republics, which lack defense treaties or guarantees of autonomy from powers capable of protecting them.

[15] After the Bolshevik Revolution, Armenians living within the Czarist empire declared their independence for a time, but later were absorbed—with Georgia and Azerbaijan—into the Transcaucasian Republic. The separate Soviet Republic of Armenia was formed in 1936.

[16] Among the peculiar problems posed by civil disorder is the wholesale theft of telephone cable, which had deprived the capital of Tbilisi of 20,000 access lines in 1993. See T. May, "The Independent States: A Survey," in *International Communications Practice Handbook*, Federal Communications Bar Association, 1993.

[17] Belorussia declared its independence from Russia after the Bolshevik Revolution, but was proclaimed a Soviet Republic in 1919. Poland acquired the western part of the territory in 1921, after fighting a brief war with the Russian Communist government, and that territory was restored to the USSR after Germany's defeat of Poland in 1939. The 1939 boundaries were confirmed at the end of the Second World War.

[18] Regulations for the telecommunications sector were adopted in 1992.

[19] The telecommunications sector had been operated and regulated by Magyar Posta (part of the ministry), but the regulatory function was transferred to the ministry's Postal and Telecommunications Inspectorate in 1989. Magyar Posta was then split into a postal agency, a broadcasting company, and a telecommunications company (HTC) in 1990.

[20] B. Wellenius and P. Stern, *Implementing Reforms in the Telecommunications Sector*, World Bank, Washington, DC, 1994, p. 380.

[21] The local concessions will be exclusive, but competitors of HTC may compete for the concessions if local governments within the 56 regions request it. If a competitor wins the concession, the service obligations of HTC in the area are at an end.

[22] Many villages, for example, are still served by manual switches that operate only during daylight hours.

[23] The boundaries of present-day Poland were set by the outcome of World War II, when lands ceded to the USSR were almost double the size of the territories acquired from defeated Germany.

[24] See K.-H. Neumann and T. Schnoring, "Telecommunications in Germany," in *Implementing Reforms in the Telecommunications Sector*, World Bank, Washington, DC, 1994, p. 327.

[25] The German government wishes to sell 49% of DBP Telekom, but this will require a change to the federal constitution.

[26] For largely political reasons, telephone service rates in the East were lowered to Western levels. Because of the scarcity of facilities in the East, this initiative reduced the effectiveness of price as a rationing mechanism during the transition to parity of service availability in the two regions.

14 Latin America

Since the mid-1980s, many of the principal economies of Latin America have moved with remarkable speed to privatize their PTTs and introduce competition in a wide range of telecommunications markets. These efforts, and the private investment needed to make them successful, have benefited from an era of democracy and political stability throughout the region.

This chapter surveys the background and present status of the liberalization process in a number of South and Central American countries, beginning with the pioneering program of the government of Chile.

14.1 CHILE

Chile boasts the most open telecommunications system in the Western Hemisphere and perhaps the most liberal in the world. The case of Chile is particularly interesting, since the economic policies of the country have covered the political spectrum over the last 20 years. Probably no other country has gone so far so fast in its trend toward liberalization.

Telephony in Chile was established in 1880 by Compañia de Teléfonos de Edison in Valparaíso. In 1927, the company was sold to International Telephone and Telegraph Corporation (ITT). It became a stock company in 1930. Until the 1980s, the Chilean telecommunications scene had more in common with that of the developing nations than with that of Europe and the United States. Telephone penetration was very low, there was limited access to international lines, and calls were priced out of reach of most people. Chile adhered to the common view that certain public utilities, including telecommunications, were best placed in the hands of the government, which was most likely to provide for universal service. There was no distinction between the government, the regulator, and the telecommunications operator. Nor was there a distinct telecommunications law. The telecommunications sector was regulated under the Electric

Utilities Law. Although the law did not require it, telecommunications was a monopoly in the hands of CTC, a company that had been owned by ITT until 1967. With 300,000 lines, CTC had 95% of the national market.

CTC became a mixed public-private enterprise in 1967 when the government invested in the company with the intention of modernizing the network. Under the government program, 80% of CTC's shares were owned by Corporactión de Fomento (CORFO), a government holding company. Three other small telephone companies—Compañia Nacional de Teléfonos, Compañia de Teléfonos de Manquehue, and Compañia de Teléfonos de Coyhaique provided local and long-distance domestic service. Both were privately owned. The government created a fifth company, Empresa Nacional de Telecomunicaciones S.A. (ENTEL), to provide long-distance facilities to CTC and the other local companies. Until very recently, ENTEL was only a carrier's carrier and the Chilean signatory to INTELSAT; subscribers and customers could not access it directly, but were billed by their local service providers for ENTEL's services.

In 1971 the Allende government nationalized telecommunications and expropriated all four companies, but by 1975 a new government began implementing a free market strategy to promote national recovery. An important part of this program was liberalization of telecommunications. Subscribers were permitted to connect their own equipment to the network and to resell their lines. Finally, a regulatory agency, the Subsecretaría de Telecomunicaciones, was established. Although the telecommunications operators were still owned by the government, they were now regulated to permit the beginnings of competition. The 1980s saw the introduction of a general telecommunications law to promote competition. In 1987, the government sold its shares in both CTC and ENTEL. Telefónica de España, the Spanish telephone company, acquired substantial interests in both companies. In 1993, the Competition Court forced Telefónica to divest itself of one of the two companies, and Telefónica elected to sell its shares in ENTEL.

In March of 1994, the Chilean parliament approved amendments to the 1982 telecommunications law. The amendments established the conditions under which the new competitive telecommunications market would operate. The new law completely liberalized the market for long-distance and international communications. A large number of foreign companies invested in the newly liberalized market by acquiring substantial interests in Chilean carriers. The new law also permitted the local telephone company, CTC, to offer long-distance and international services through an affiliate, CTC Mundo. But the most important change was the introduction of the *multicarrier system*, which allows every caller to choose the carrier that will terminate his or her long-distance call. The new law also authorized resale and mandated dialing parity and equal interconnection terms for all carriers.

The multicarrier system was implemented in phases. The country was divided into regions, and a deadline was selected for each. The last region in which the new system was implemented was the Santiago metropolitan area, where the system became effective on October 29, 1994.

As a result of the liberalization process, Chile today can boast a modern telecommunications system and more telecommunications companies per capita than any other nation in the world. Although the market is small by world standards (1.6 million lines, with revenues of US$220 million), the carriers have been exceptionally aggressive in their marketing efforts. A savage price war followed the introduction of competition with prices of calls to the United States dipping to as little as 1 cent per minute.

Chile has also been a pioneer of wireless technologies in Latin America. It was one of the first countries to introduce cellular technology in 1989, and cellular service soon became a popular alternative to the spotty coverage of the wireline company. Chile will also be one of the first countries to introduce PCS. In April of 1995, the government published the terms of reference for the Chilean regulations that will govern PCS and invited public comment. Three PCS licenses will be awarded.

Chile represents a unique model in telecommunications. No other country has undergone such a drastic liberalization in so short a time. Although the process has attracted investors and has been responsible for the birth of new companies, the jury is still out on whether the Chilean model is the best way to manage the transition from a state-controlled monopoly to full competition.

14.2 ARGENTINA

Before the recent liberalization, most telecommunications services in Argentina were provided by Empresa Nacional de Telecomunicaciones (ENTel)—a state-owned monopoly formed in the 1940s [1]. Under ENTel, the penetration rate was only about 8.8% and the waiting time for service averaged over four years.

Liberalization in Argentina began with a 1989 statute that authorized privatization of state-owned enterprises generally, and directed that ENTel would be privatized, in particular. The statute, called the State Reform Law, gave the executive the power to carry out privatizations and required that they be accomplished through competitive bidding.

The executive's first task in implementing the statute was to prepare for privatization by defining, for the benefit of potential bidders, the environment in which the privatized company—or companies—would operate. To assist with this task, the executive assembled a privatization group that included a number of foreign and domestic consultants.

Eventually, it was decided that two corporations would be created out of ENTel, one to serve northern Argentina and the other to serve southern Argentina. (Each company would also serve a portion of the Buenos Aires metropolitan area.) The new companies would enjoy monopolies of basic services in their geographic areas for seven years, with a possible extension of three additional years if prescribed performance goals were met. The monopoly franchise would then be followed by a regime of open competition. International and nonbasic domestic services [2] would be provided by two additional companies, both of which would be owned by the two regional monopoly carriers [3].

The program also called for successful bidders to be assured of a rate of return on investment sufficient to support the service obligations to which the regional companies would be subject. Specifically, during an initial transition period of two years after privatization, the owners could request tariff adjustments sufficient to provide a rate of return of up to 16%. After the transition period, a price cap regime would take effect, and the companies had to reduce their rates 2% (adjusted for inflation) during each remaining year of the monopoly franchise [4].

The privatization was accomplished in two phases. First, interested bidders established their eligibility under certain criteria, including net worth and operating experience. Entities that survived this prequalifying process were then furnished with complete information concerning the condition of ENTel and the terms of the proposed concessions, and were invited to bid. Each bidder was to offer a mandatory cash amount and state the share of ENTel's debt that it was prepared to assume. Foreign companies could own as much as 40% of each regional company, and the bidding was to be managed so that each regional company would be under different ownership.

In spite of the complexity of this procedure, the sale of ENTel's successor entities was completed almost on schedule. By the end of 1990, Argentina had two private telephone companies: Telecom Argentina in the north, owned by a group that included France Télécom and STET of Italy, and Telefonica de Argentina in the south, owned by a consortium headed by Telefónica de España. (10% of the shares in the two companies were sold to the public and 30% to employees.)

While the privatization effort in Argentina has been a success, the adoption of an adequate regulatory regime has faltered almost from the start. While the executive established a telecommunications regulatory agency—Comision Nacional do Telecomunicaciones (CNT)—in 1990, the agency had weak political support and was the creature of a decree that could be revoked at any time. CNT has been starved for funds, poorly staffed, and has undermined investor confidence by putting pressure on the regional carriers to lower their tariffs. Finally, in 1995, the executive seized control of CNT, effectively putting it out of business.

In spite of these difficulties, the telecommunications sector in Argentina has performed reasonably well and investors have enjoyed strong returns from operations. Whether the regional companies' monopolies will be extended when the initial franchise period expires in 1997, however, remains to be seen.

14.3 VENEZUELA

Until recently, telephone service in Venezuela was provided by a state-owned monopoly called Compania Anonima Nacional do Telefonos de Venezuela (CANTV). The penetration rate was 8.2%, and the waiting time for service was about 18 months.

The decision to privatize CANTV was made in 1990, and in 1991 the government proposed a two-step privatization process somewhat similar to Argentina's. Bidders were required to satisfy a prequalification process, after which detailed bidding requirements were provided to the winners. The two qualified consortia—one led by Bell Atlantic and the other led by GTE—competed in the second round and the GTE group prevailed.

The terms of the GTE group's concession include a nine-year monopoly of local, long-distance, and international service. During this period, the company is required to meet exacting service and network expansion targets. As an additional incentive to meet these goals, the company was placed under a price cap system of price regulation.

Venezuela, like Argentina, regulates its telecommunications sector through an agency created by executive decree rather than legislation. This agency, called Consejo Nacional de Telecomunicaciones, is within the Ministry of Transport and Communications.

14.4 BRAZIL

Brazil, with one of the largest telephone networks in the world, has not privatized its system in spite of chronic service problems and substantial unmet demand. Most local services continue to be provided by Telecomunicacoes Brasileiras S.A., a government holding company that controls 29 operating companies. Long-distance and international services are provided by another state monopoly called Empresa Brasileira de Telecomunicacoes.

While so far refusing to sell these enterprises or open basic services to competition, the Brazilian government has permitted some liberalization of nonbasic markets. Notably, decrees of the Ministry of Infrastructure have permitted private networks, competing provision of value-added services, and competition in satellite and cellular service. And, most recently, in September

1995, Brazil's Congress opened the way to privatization by amending the constitution to permit private investment in telecommunications companies.

14.5 BOLIVIA

Bolivia's telephone penetration rate of 2.7% is among the lowest in Latin America, and the need to privatize the government monopoly (ENTEL) and attract new capital has been obvious for many years. Nevertheless, the government has lost much time pursuing halfway measures that have met with only isolated success.

Telephone service in Bolivia is fragmented, with some 17 cooperatives providing local service and ENTEL providing long-distance and international service. The cooperative structure, which is a poor device for attracting new capital, was made ubiquitous in the local service sector by a government initiative undertaken in 1985. More recently, a governmental commission has recommended that ENTEL be privatized and the local cooperatives be reorganized as joint stock companies. The government, which hopes to improve the penetration rate to 5.2% by the year 2000, is inclined to follow these recommendations.

14.6 COLOMBIA

Colombia, like Bolivia, is served by a state monopoly provider of international and long-distance service and a large number of private local carriers. The state monopoly—Empresa Nacional de Telecomunicaciones (TELECOM)—also owns a number of local telephone companies.

Because of the fragmentation of the industry and the variety of economic conditions prevailing in different regions of the country, the quality and availability of telephone service vary widely. Some of the large urban companies are quite efficient, while rural service tends to be quite poor.

The government has decided in principle to privatize TELECOM and to introduce competition (at least in long distance), but little concrete progress toward liberalization of the sector has been made.

14.7 PERU

Peru, with one of the poorest telephone networks in the region, historically has been served by two telephone companies: Empresa Nacional de Telecomunicaciones (ENTEL), which has provided international and long-distance service on a monopoly basis, as well as local service to many areas, and Compañía Peruana de Teléfonos (CPT), which has provided monopoly service to the Lima area. CPT was owned primarily by its subscribers, with the government holding a 22% share, and ENTEL was entirely government-owned.

ENEL and CPT have since been privatized under a program devised by the Comité de Privatización (CPRI), a governmental committee set up in 1992 to plan and oversee the divestiture of all state-owned enterprises. The terms and conditions under which competition will be permitted and the sector will be regulated are still under development.

14.8 ECUADOR

All telecommunications services in Ecuador are still provided by Empresa Estatal de Telcomunicaciones (EMETEL), a quasi-independent state-owned enterprise. Ecuador has, however, taken a number of steps that will facilitate the eventual privatization of EMETEL. Notably, the government passed a new telecommunications law in 1992, which gave EMETEL control over its own budget and revenues and established an independent telecommunications regulator, the Superintendencia de Telcomunicaciones.

While the government has announced its long-range intention to privatize EMETEL, the present emphasis is on further corporatization and streamlining of the entity. This effort is under the general supervision of the Consejo de Modernización del Estado, a governmental agency created to oversee the improvement of the performance of the public sector in Ecuador.

14.9 URUGUAY

Uruguay is served by a state telecommunications monopoly called Administración Nacional de Telcomunicaciones (ANTEL), which provides relatively efficient service to the Montivideo area, but operates a severely antiquated system in the countryside. The telecommunications sector is regulated by the National Department of Communications, an organ of the Ministry of Defense, and entities other than ANTEL are permitted to provide cellular, value-added, and other nonbasic services. (Local, long-distance, and international services remain an ANTEL monopoly.)

Privatization of ANTEL appeared to be assured after the government, acting pursuant to expert advice and under public sector reform laws passed by the Congress in 1990, recommended that ANTEL be partially sold. Under the government recommendation, made in 1991, 49% of the equity of the company would be in private hands, the government would retain a 51% share, and foreign ownership would be limited to 49% of the total. Those plans were put on hold, however, when a referendum held in December 1992 rejected the privatization plan. The result of this referendum also prevented the planned establishment of a new independent telecommunications regulatory commission.

Notes

[1] ENTel was formed by nationalization of a private carrier owned by ITT. In addition to the ITT-owned carrier, six of Argentina's 22 provinces were served by a separate private company owned by L. M. Ericsson. The latter company—called CAT—was still privately owned when liberalization of the telecommunications sector began.

[2] Certain services, including private networks, customer premises equipment and value-added services, were to be opened to full competition.

[3] Specifically, assets used by ENTel to provide international and competitive services, respectively, were transferred to the new corporations, and shares in those corporations were retained by the two regional monopolies.

[4] If the three-year extension of the monopoly franchise was granted, the companies were required to achieve reductions of 4% annually during those three years.

Appendix A

The following table is a portion of the FCC's version of the Table of Frequency Allocations, which shows both international allocations and U.S. variants. (*Source:* U.S. Government Printing Office, 47 Code of Federal Regulations, Section 2.106).

International table			United States table		FCC use designators	
Region 1—allocation MHz	Region 2—allocation MHz	Region 3—allocation MHz	Government	Non-Government	Rule part(s)	Special-use frequencies
			Allocation MHz	Allocation MHz		
(1)	(2)	(3)	(4)	(5)	(6)	(7)
FIXED. MOBILE except aeronautical mobile. 722	FIXED. MOBILE 723. 722		FIXED. MOBILE. 722 G30	Land Mobile (telemetering and telecommand). Fixed (telemetering) 722	Private Land Mobile (90).	
			1435–1530 MOBILE (aeronautical telemetering)	1435–1530 MOBILE (aeronautical telemetering)	AVIATION (87).	
1525–1530 SPACE OPERATION (space-to-Earth). FIXED. MARITIME MOBILE–SATELLITE (space-to-Earth). Land Mobile–Satellite (space-to-Earth) 726B. Earth Exploration-Satellite Mobile except aeronautical mobile 724. 722 723B 725 726A 726D	1525–1530 SPACE OPERATION (space-to-Earth). MOBILE–SATELLITE (space-to-Earth). Earth Exploration–Satellite. Fixed. Mobile 723. 722 723A 726A 726D	1525–1530 SPACE OPERATION (space-to-Earth). FIXED. MOBILE–SATELLITE (space-to-Earth). Earth Exploration–Satellite. Mobile 723 724. 722 726A 726D	722 US78	722 US78		
1530–1533 SPACE OPERATION (space-to-Earth). MARITIME MOBILE–SATELLITE (space-to-Earth). LAND MOBILE-SATELLITE (space-to-Earth). Earth Exploration–Satellite. Fixed. Mobile except aeronautical mobile.	1530–1533 SPACE OPERATION (space-to-Earth). MARITIME MOBILE–SATELLITE (space-to-Earth). LAND MOBILE–SATELLITE (space-to-Earth). Earth Exploration-Satellite. Fixed. Mobile 723	1530–1533 SPACE OPERATION (space-to-Earth). MARITIME MOBILE–SATELLITE (space-to-Earth). LAND MOBILE–SATELLITE (space-to-Earth). Earth Exploration- - Satellite. Fixed. Mobile 723	1530–1535 MARITIME MOBILE–SATELLITE (space-to-Earth). MOBILE–SATELLITE (space-to-Earth). Mobile (aeronautical telemetering).	1530–1535 MARITIME MOBILE–SATELLITE (space-to-Earth). MOBILE–SATELLITE (space-to-Earth). Mobile (aeronautical telemetering).	SATELLITE COMMUNICATION (25). Aviation (87).	

722 723B 726A 726D	722 726A 726C 726D	722 726A 726C 726D				
1533–1535 SPACE OPERATION (space-to-Earth). MARITIME MOBILE–SATELLITE (space-to-Earth). Earth Exploration-Satellite. Fixed. Mobile except aeronautical mobile. Land Mobile–Satellite (space-to-Earth) 726B. 722 723B 726A 726D	1533–1535 SPACE OPERATION (space-to-Earth). MARITIME MOBILE–SATELLITE (space-to-Earth). Earth Exploration–Satellite. Fixed. Mobile 723. Land Mobile–Satellite (space-to-Earth) 726B. 722 726A 726C 726D	1533–1535 SPACE OPERATION (space-to-Earth). MARITIME MOBILE–SATELLITE (space-to-Earth). Earth Exploration–Satellite. Fixed. Mobile 723. Land Mobile–Satellite (space-to-Earth) 726B. 722 726A 726C 726D	722 726A US78 US315	722 726A US78 US315		
1544–1545 MOBILE–SATELLITE (space-to-Earth). 722 726D 727 727A	1544–1545 MOBILE–SATELLITE (space-to-Earth). 722 726D 727 727A	1544–1545 MOBILE–SATELLITE (space-to-Earth). 722 726D 727 727A	1544–1545 MOBILE–SATELLITE (space-to-Earth). 722 727A	1544–1545 MOBILE–SATELLITE (space-to-Earth). 722 727A	MARITIME (80). SATELLITE COMMUNICATION (25).	
1545–1555 AERONAUTICAL MOBILE–SATELLITE (R) (space-to-Earth). 722 726A 726D 727 729 729A 730	1545–1555 AERONAUTICAL MOBILE–SATELLITE (R) (space-to-Earth). 722 726A 726D 727 729 729A 730	1545–1555 AERONAUTICAL MOBILE–SATELLITE (R) (space-to-Earth). 722 726A 726D 727 729 729A 730	1545–1549.5 AERONAUTICAL MOBILE–SATELLITE (R) (space-to-Earth). Mobile–satellite (space-to-Earth). 722 726A US308 US309	1545–1549.5 AERONAUTICAL MOBILE–SATELLITE (R) (space-to-Earth). Mobile–satellite (space-to-Earth). 722 726A US308 US309	AVIATION (87)	
			1549.5–1558.5 AERONAUTICAL MOBILE–SATELLITE (R) (space-to-Earth) MOBILE–SATELLITE (space-to-Earth). 722 726A US308 US309	1549.5–1558.5 AERONAUTICAL MOBILE–SATELLITE (R) (space-to-Earth). MOBILE–SATELLITE (space-to-Earth). 722 726A US308 US309	AVIATION (87).	
1555–1559 LAND MOBILE–SATELLITE (space-to-Earth). 722 726A 726D 727 730 730A 730B 730C	1555–1559 LAND MOBILE–SATELLITE (space-to-Earth). 722 726A 726D 727 730 730A 730B 730C	1555–1559 LAND MOBILE–SATELLITE (space-to-Earth). 722 726A 726D 727 730 730A 730B 730C	1558.5–1559 AERONAUTICAL MOBILE–SATELLITE (R) (space-to-Earth). 722 726A US308 US309	1558.5–1559 AERONAUTICAL MOBILE–SATELLITE (R) (space-to-Earth). 722 726A US308 US309	AVIATION (87).	

International table			United States table		FCC use designators	
Region 1—allocation MHz	Region 2—allocation MHz	Region 3—allocation MHz	Government	Non-Government	Rule part(s)	Special-use frequencies
			Allocation MHz	Allocation MHz		
(1)	(2)	(3)	(4)	(5)	(6)	(7)
1559–1610	AERONAUTICAL RADIONAVIGATION. RADIONAVIGATION-SATELLITE (space-to-Earth). 722 727 730 731 731A 731B 731C 731D		1559–1610 AERONAUTICAL RADIONAVIGATION-SATELLITE (space-to-Earth). 722 US208 US260	1559–1610 AERONAUTICAL RADIONAVIGATION-SATELLITE (space-to-Earth). 722 US208 US260	AVIATION (87).	
1610–1610.6 AERONAUTICAL RADIO-NAVIGATION. MOBILE-SATELLITE (Earth-to-space). 722 727 730 731 731E 732 733 733A 733B 733E 733F	1610–1610.6 AERONAUTICAL RADIONAVIGATION. RADIO-DETERMIN-ATION SATELLITE (Earth-to-space). MOBILE-SATELLITE (Earth-to-space). 722 731E 732 733 733A 733C 733D 733E	1610–1610.6 AERONAUTICAL RADIONAVIGATION. MOBILE-SATELLITE (Earth-to-space). Radiodetermination-Satellite(Earth-to-space). 722 727 730 731E 732 733 733A 733B 733E	1610–1610.6 AERONAUTICAL RADIONAVIGATION. RADIODETERMIN-ATION SATELLITE (Earth-to-space). MOBILE-SATELLITE (Earth-to-space). 722 731E 732 733 733A 733E US208 US260 US319	1610–1610.6 AERONAUTICAL RADIONAVIGATION. RADIODETERMIN-ATION SATELLITE (Earth-to-space). MOBILE-SATELLITE (Earth-to-space). 722 731E 732 733 733A 733E US208 US260 US319	AVIATION (87). SAT-ELLITE COMMUN-ICATION (25).	
1610.6–1613.8 AERONAUTICAL RADIO-NAVIGATION. MOBILE-SATELLITE (Earth-to-space). RADIO-ASTRONOMY 722 727 730 731 731E 732 733 733A 733B 733E 733F 734	1610.6–1613.8 AERONAUTICAL RADIONAVIGATION. RADIODETERMIN-ATION SATELLITE (Earth-to-space). MOBILE-SATELLITE (Earth-to-space). RADIO-ASTRONOMY 722 731E 732 733 733A 733C 733D 733E 734	1610.6–1613.8 AERONAUTICAL RADIONAVIGATION. MOBILE-SATELLITE (Earth-to-space). RADIO-ASTRONOMY. RADIODETERMIN-ATION SATELLITE (Earth-to-space). 722 727 730 731E 732 733 733A 733B 733E 734	1610.6–1613.8 AERONAUTICAL RADIONAVIGATION. RADIODETERMIN-ATION SATELLITE (Earth-to-space). MOBILE-SATELLITE (Earth-to-space). RADIO-ASTRONOMY 722 731E 732 733 733A 733E 734 US208 US260 US319	1610.6–1613.8 AERONAUTICAL RADIONAVIGATION. RADIODETERMIN-ATION SATELLITE (Earth-to-space). RADIO-ASTRONOMY MOBILE-SATELLITE (Earth-to-space). 722 731E 732 733 733A 733E 734 US208 US260 US319	AVIATION (87). SAT-ELLITE COMMUN-ICATION (25).	

1613.8–1626.5 AERONAUTICAL RADIO-NAVIGATION. MOBILE-SATELLITE (Earth-to-space). Mobile-Satellite (space-to-Earth). 722 727 730 731 731E 731F 732 733 733A 733B 733E 733F	1613.8–1626.5 AERONAUTICAL RADIONAVIGATION. RADIODETERMIN-ATION SATELLITE (Earth-to-space). MOBILE-SATELLITE (Earth-to-space). Mobile-Satellite (space-to-Earth). 722 731E 731F 732 733 733A 733C 733D 733E	1613.8–1626.5 AERONAUTICAL RADIONAVIGATION. MOBILE-SATELLITE (Earth-to-space). Radiodetermination Satellite (Earth-to-space). Mobile-Satellite (space-to-Earth). 722 727 730 731E 731F 732 733 733A 733B 733E	1613.8–1626.5 AERONAUTICAL RADIONAVIGATION. RADIODETERMIN-ATION SATELLITE (Earth-to-space). MOBILE-SATELLITE (Earth-to-space). Mobile-Satellite (space-to-Earth). 722 731E 731F 732 733 733E US208 US260 US319	1613.8–1626.5 AERONAUTICAL RADIONAVIGATION. RADIODETERMIN-ATION SATELLITE (Earth-to-space). MOBILE-SATELLITE (Earth-to-space). Mobile-Satellite (space-to-Earth). 722 731E 731F 732 733 733E US208 US260 US319	AVIATION (87). SAT-ELLITE COMMUN-ICATION (25).	
1626.5–1631.5 MARITIME MOBILE-SAT-ELLITE (Earth-to-space). Land Mobile-Satellite (Earth-to-space) 726B. 722 726A 726D 727 730	1626.5–1631.5 MOBILE-SATELLITE (Earth-to-space). 722 726A 726C 726D 727 730	1626.5–1631.5 MOBILE-SATELLITE (Earth-to-space). 722 726A 726C 726D 727 730	1626.5–1645.5 MARITIME MOBILE-SATELLITE (Earth-to-space). MOBILE-SATELLITE (Earth-to-space).	1626.5–1645.5 MARITIME MOBILE-SATELLITE (Earth-to-space). MOBILE-SATELLITE (Earth-to-space).	MARITIME (80). SATELLITE COMMUN-ICATION (25).	
1631.5–1634.5 MARITIME MOBILE-SAT-ELLITE (Earth-to-space). LAND MOBILE-SAT-ELLITE (Earth-to-space). 722 726A 726C 726D 727 730 734A	1631.5–1634.5 MARITIME MOBILE-SATELLITE (Earth-to-space). LAND MOBILE-SAT-ELLITE (Earth-to-space). 722 726A 726C 726D 727 730 734A	1631.5–1634.5 MARITIME MOBILE-SATELLITE (Earth-to-space). LAND MOBILE-SAT-ELLITE (Earth-to-space). 722 726A 726C 726D 727 730 734A				
1634.5–1645.5 MARITIME MOBILE-SAT-ELLITE (Earth-to-space). Land Mobile-Satellite (Earth-to-space) 726B. 722 726A 726C 726D 727 730	1634.5–1645.5 MARITIME MOBILE-SATELLITE (Earth-to-space). Land Mobile-Satellite (Earth-to-space) 726B. 722 726A 726C 726D 727 730	1634.5–1645.5 MARITIME MOBILE-SATELLITE (Earth-to-space). Land Mobile-Satellite (Earth-to-space) 726B. 722 726A 726C 726D 727 730	722 726A US315	722 726A US315		
1645.5–1646.5 MOBILE-SATELLITE (Earth-to-space). 722 726D 734B	1645.5–1646.5 MOBILE-SATELLITE (Earth-to-space). 722 726D 734B	1645.5–1646.5 MOBILE-SATELLITE (Earth-to-space). 722 726D 734B	1645.5–1646.5 MOBILE-SATELLITE (Earth-to-space). 722 734B	1645.5–1646.5 MOBILE-SATELLITE (Earth-to-space). 722 734B	MARITIME (80). SATELLITE COMMUN-ICATION (25).	

International table			United States table		FCC use designators	
Region 1—allocation MHz	Region 2—allocation MHz	Region 3—allocation MHz	Government	Non-Government	Rule part(s)	Special-use frequencies
			Allocation MHz	Allocation MHz		
(1)	(2)	(3)	(4)	(5)	(6)	(7)
1646.5–1656.5 AERONAUTICAL MOBILE-SATELLITE (R) (Earth-to-space). 722 726A 726D 727 729A 730 735	1646.5–1656.5 AERONAUTICAL MOBILE-SATELLITE (R) (Earth-to-space). 722 726A 726D 727 729A 730 735	1646.5–1656.5 AERONAUTICAL MOBILE-SATELLITE (R) (Earth-to-space). 722 726A 726D 727 729A 730 735	1646.5–1651 AERONAUTICAL MOBILE-SATELLITE (R) (Earth-to-space) Mobile-Satellite (Earth-to-space).	1646.5–1651 AERONAUTICAL MOBILE-SATELLITE (R) (Earth-to-space) Mobile-Satellite (Earth-to-space). 722 726A US308 US309	AVIATION (87). 722 726A US308 US309	
1656.5–1660 LAND MOBILE-SATELLITE (Earth-to-space). 722 726A 726D 727 730 730A 730B 730C 734A	1656.5–1660 LAND MOBILE-SATELLITE (Earth-to-space). 722 726A 726D 727 730 730A 730B 730C 734A	1656.5–1660 LAND MOBILE-SATELLITE (Earth-to-space). 722 726A 726D 727 730 730A 730B 730C 734A	1651–1660 AERONAUTICAL MOBILE-SATELLITE (R) (Earth-to-space) MOBILE-SATELLITE (Earth-to-space). 722 726A US308 US39	1651–1660 AERONAUTICAL MOBILE-SATELLITE (R) (Earth-to-space) MOBILE-SATELLITE (Earth-to-space). 722 726A US308 US39	AVIATION (87).	
1660–1660.5 RADIO-ASTRONOMY LAND MOBILE-SATELLITE (Earth-to-space). 722 726A 726D 730A 730B 730C 736	1660–1660.5 RADIO-ASTRONOMY LAND MOBILE-SATELLITE (Earth-to-space). 722 726A 726D 730A 730B 730C 736	1660–1660.5 RADIO-ASTRONOMY LAND MOBILE-SATELLITE (Earth-to-space). 722 726A 726D 730A 730B 730C 736	1660–1660.5 AERONAUTICAL MOBILE-SATELLITE (R) (Earth-to-space) RADIO ASTRONOMY. 722 726A 736 US309	1660–1660.5 AERONAUTICAL MOBILE-SATELLITE (R) (Earth-to-space) RADIO ASTRONOMY. 722 726A 736 US309	AVIATION (87).	
1660.5–1668.4	RADIO ASTRONOMY. SPACE RESEARCH (passive). Fixed. Mobile except aeronautical mobile. 722 736 737 738 739		1660.5–1668.4 RADIO ASTRONOMY. SPACE RESEARCH (passive). 722 US74 US246	1660.5–1668.4 RADIO ASTRONOMY. SPACE RESEARCH (passive). 722 US74 US246		
1668.4–1670			1668.4–1670.0	1668.4–1670.0		

	METEOROLOGICAL AIDS. FIXED. MOBILE except aeronautical mobile. RADIO ASTRONOMY 722 736		METEOROLOGICAL AIDS (radiosonde). RADIO ASTRONOMY 722 736 US74 US99	METEOROLOGICAL AIDS (radiosonde). RADIO ASTRONOMY 722 736 US74 US99		
1670–1675	METEOROLOGICAL AIDS. FIXED. METEOROLOGICAL-SATELLITE (space-to-Earth). MOBILE 740A 722		1670–1690 METEROLOGICAL AIDS (radiosonde) METEOROLOGICAL-SATELLITE (space-to-Earth).	1670–1690 METEROLOGICAL AIDS (radiosonde) METEOROLOGICAL-SATELLITE (space-to-Earth).		
1675–1690 METEOROLOGICAL AIDS. FIXED. METEOROLOGIAL-SATELLITE (space-to-Earth). MOBILE except aeronautical mobile. 722	1675–1690 METEOROLOGICAL AIDS. FIXED. METEOROLOGIAL-SATELLITE (space-to-Earth). MOBILE except aeronautical mobile. 722 735A	1675–1690 METEOROLOGICAL AIDS. FIXED. METEOROLOGIAL-SATELLITE (space-to-Earth). MOBILE except aeronautical mobile. 722	722 US211	722 US211		
1690–1700 METEOROLOGICAL AIDS. METEOROLOGIAL-SATELLITE (space-to-Earth). FIXED. MOBILE except aeronautical mobile. 671 722 741	1690–1700 METEOROLOGICAL AIDS. METEOROLOGIAL SATELLITE (space-to-Earth). MOBILE-SATELLITE (Earth-to-space). 671 722 735A 740	1690–1700 METEOROLOGICAL AIDS. METEOROLOGIAL SATELLITE (space-to-Earth). 671 722 740 742	1690–1700 METEOROLOGICAL AIDS (Radiosonde) METEOROLOGIAL SATELLITE (space-to-Earth). 671 722	1690–1700 METEOROLOGICAL AIDS (Radiosonde) METEOROLOGIAL SATELLITE (space-to-Earth). 671 722		
1700–1710 FIXED. METEOROLOGICAL-SATELLITE (space-to-Earth). MOBILE except aeronautical mobile. 671 722	1700–1710 FIXED. METEOROLOGICAL-SATELLITE (space-to-Earth). MOBILE except aeronautical mobile. MOBILE-SATELLITE (Earth-to-space). 671 722 735A	1700–1710 FIXED. METEOROLOGICAL-SATELLITE (space-to-Earth). MOBILE except aeronautical mobile. 671 722 743	1700–1710 FIXED. METEOROLOGICAL-SATELLITE (space-to-Earth). 671 722	1700–1710 METEOROLOGICAL-SATELLITE (space-to-Earth). Fixed. 671 722		

International table			United States table		FCC use designators	
Region 1—allocation MHz	Region 2—allocation MHz	Region 3—allocation MHz	Government	Non-Government	Rule part(s)	Special-use frequencies
			Allocation MHz	Allocation MHz		
(1)	(2)	(3)	(4)	(5)	(6)	(7)
			G118			
1710–1930	FIXED. MOBILE 740A 722 744 745 746 746A		1710–1850 FIXED. MOBILE. 722 US256 G42	1710–1850 722 US256		
1930–1970 FIXED. MOBILE. 746A	1930–1970 FIXED. MOBILE MOBILE-SATELLITE (Earth-to-space). 746A	1930–1970 FIXED. MOBILE. 746A	1850–1990	1850–1990 FIXED MOBILE US331	Personal communications services (24) Private operational-fixed microwave (94) Radio frequency devices (15)	EMERGING TECHNOLOGIES.
1970–1980 FIXED. MOBILE. 746A	1970–1980 FIXED. MOBILE. MOBILE-SATELLITE (Earth-to-space). 746A 746B 746C	1970–1980 FIXED. MOBILE. 746A				
1980–2010	FIXED. MOBILE. MOBLE-SATELLITE (Earth-to-space). 746A 746B 746C		1990–2110	1990–2110 FIXED. MOBILE	AUXILIARY BROADCAST (74). CABLE TELEVISION (78).	
2010–2025	FIXED. MOBILE. 746A					
2025–2110	FIXED. MOBILE 747A.					

	SPACE RESEARCH (Earth-to-space) (space-to-space) SPACE OPERATION (Earth-to-space) (space-to-space) EARTH EXPLOR- ATION-SATELLITE (Earth-to-space) (space-to-space). 750A		US90 US111 US219 US222	US90 US111 US219 US222 NG23 NG118		
2110–2120	FIXED. MOBILE. SPACE RESEARCH (deep space) (Earth- to-space). 746A		2100–2200	2110–2150 FIXED MOBILE US111 US252 NG23 NG153	Domestic public fixed (21) Private operational fixed microwave (94) Public mobile (22)	EMERGING TECH- NOLOGIES
2120–2160 FIXED. MOBILE. 746A	2120–2160 FIXED. MOBILE. MOBILE-SATELLITE (space-to-Earth). 746A	2120–2160 FIXED. MOBILE. 746A		2150–2160 FIXED NG23	Multipoint distribution (21) Private operational-fixed microwave (94)	
2160–2170 FIXED. MOBILE. 746A	2160–2170 FIXED. MOBILE. MOBILE-SATELLITE (space-to-Earth). 746A 746B 746C	2160–2170 FIXED. MOBILE. 746A		2160–2200 FIXED MOBILE	Domestic public fixed (21) Private operational-fixed microwave (94) Public mobile (22)	EMERGING TECH- NOLOGIES
2170–2200	FIXED. MOBILE. MOBILE-SATELLITE (space-to-Earth). 746A 746B 746C		US111 US252 US331	NG23 NG153		
2200–2290	FIXED.		2200–2290	2200–2290		

International table			United States table		FCC use designators	
Region 1—allocation MHz	Region 2—allocation MHz	Region 3—allocation MHz	Government	Non-Government	Rule part(s)	Special-use frequencies
			Allocation MHz	Allocation MHz		
(1)	(2)	(3)	(4)	(5)	(6)	(7)
	SPACE RESEARCH (space-to-Earth) (space-to-space). SPACE OPERATION (space-to-Earth) (space-to-space) EARTH EXPLOR- ATION-SATELLITE (space-to-Earth) (space-to-space) MOBILE 747A 750A		FIXED. MOBILE. SPACE RESEARCH. (Space-to-Earth) (Space-to-space) US303 G101	US303		
2290–2300 FIXED. SPACE RESEARCH (space-to-Earth) (deep space). MOBILE except aero-naut- ical mobile. 743A	2290–2300 FIXED. MOBILE except aero- nautical mobile. SPACE RESEARCH (space-to-Earth) (deep space).		2290–2300 FIXED. MOBILE except aero- nautical mobile. SPACE RESEARCH (space-to-Earth) (deep space only).	2290–2300 SPACE RESEARCH (space-to-Earth) (deep space only).		
2300–2450 FIXED. Amateur. Mobile. Radiolocation. 664 743A 752	2300–2450 FIXED. MOBILE. RADIOLOCATION. Amateur. 664 751 752		2300–2310 RADIOLOCATION Fixed. Mobile. US253 G2	2300–2310 Amateur. US253	Amateur (97).	
			2310–2390 MOBILE. RADIOLOCATION. Fixed. US276 G2	2310–2390 MOBILE US276		
			2390–2450. RADIOLOCATION 664 752 G2	2390–2450 Amateur. 664 752	Amateur (97).	
2450–2483.5	2450–2483.5		2450–2483.5	2450–2483.5		

FIXED. MOBILE. Radiolocation. 752 753	FIXED. MOBILE. RADIOLOCATION. 752		752 US41	FIXED. MOBILE. Radiolocation. 752 US41		2450 ± 50 MHz: Industrial, scientific, and medical frequency.
2483.5–2500 FIXED. MOBILE. MOBILE-SATELLITE (space-to-Earth). Radiolocation. 733F 752 753 753A 753B 753C 753F	2483.5–2500 FIXED. MOBILE. RADIODETERMINATION SATELLITE (space-to-Earth) 753A RADIOLOCATION. MOBILE-SATELLITE (space-to-Earth). 752 753D 753F	2483.5–2500 FIXED. MOBILE. RADIOLOCATION. MOBILE-SATELLITE (space-to-Earth). Radiodetermination-Satellite (space-to-Earth). 753A. 752 753C 753F	2483.5–2500 RADIODETERMINATION SATELLITE (space-to-Earth) 753A. MOBILE-SATELLITE (space-to-Earth). 752 753F US41 US319	2483.5–2500 RADIODETERMINATION SATELLITE (space-to-Earth). 753A. MOBILE-SATELLITE (space-to-Earth). 752 753F US41 US319 NG147	SATELLITE COMMUNICATION (25).	
2500–2655 FIXED 762 763 764 MOBILE except aeronautical mobile. BROADCASTING-SATELLITE 757 760 720 753 756 758 759	2500–2655 FIXED 762 764 FIXED-SATELLITE (space-to-Earth) 761 MOBILE except aeronautical mobile. BROADCASTING-SATELLITE 757 760 720 755	2500–2535 FIXED 762 764 FIXED SATELLITE (space-to-Earth) 761 MOBILE except aeronautical mobile. BROADCASTING-SATELLITE 757 760 754 754A 2535–2655 FIXED 762 764 MOBILE except aeronautical mobile. BROADCASTING-SATELLITE 757 760 720	2500–2655 720 US205 US269	2500–2655 FIXED. BROADCASTING-SATELLITE. 720 US205 US269 NG101 NG102	AUXILIARY BROADCASTING (74) DOMESTIC PUBLIC FIXED RADIO (21)	
2655–2690	2655–2690	2655–2690	2655–2690	2655–2690		

International table			United States table		FCC use designators	
Region 1—allocation MHz	Region 2—allocation MHz	Region 3—allocation MHz	Government	Non-Government	Rule part(s)	Special-use frequencies
			Allocation MHz	Allocation MHz		
(1)	(2)	(3)	(4)	(5)	(6)	(7)
FIXED 762 763 764 MOBILE except aeronautical mobile. BROADCASTING-SATELLITE 757 760 Earth Exploration-Satellite (passive) Radio Astronomy. Space Research (passive). 758 759 765	FIXED 762 764 FIXED-SATELLITE (Earth-to-space) (space-to-Earth) 761 MOBILE except aeronautical mobile. BROADCASTING-SATELLITE 757 760 Earth Exploration-Satellite (passive). Radio Astronomy. Space Research (passive). 765	FIXED 762 764 FIXED-SATELLITE (Earth-to-space) 761 MOBILE except aeronautical mobile. BROADCASTING-SATELLITE 757 760 Earth Exploration-Satellite (passive). Radio Astronomy. Space Research (passive). 765 766	Earth Exploration-Satellite (passive). Radio Astronomy. Space Research (passive). US205 US269	FIXED. BROADCASTING-SATELLITE. Earth Exploration-Satellite (passive). Radio Astronomy. Space Research (passive). US205 US269 NG47 NG101 NG102	AUXILIARY BROADCASTING (74). PRIVATE OPERATIONAL-FIXED MICROWAVE (94).	
2690–2700	EARTH EXPLORATION-SATELLITE (passive). RADIO ASTRONOMY. SPACE RESEARCH (passive). 767 768 769		2690–2700 EARTH EXPLORATION-SATELLITE (passive). RADIO ASTRONOMY. SPACE RESEARCH (passive). US74 US246	2690–2700 EARTH EXPLORATION-SATELLITE (passive). RADIO ASTRONOMY. SPACE RESEARCH (passive). US74 US246		
2700–2900	AERONAUTICAL RADIONAVIGATION 717. Radiolocation. 770 771		2700–2900 AERONAUTICAL RADIONAVIGATION 717. METEOROLOGICAL AIDS. Radiolocation. 770 US18 G2 G15	2700–2900 717 770 US18		
2900–3100	RADIONAVIGATION. Radiolocation. 772 773 775A		2900–3100 MARITIME RADIONAVIGATION 775A Radiolocation.	2900–3100 MARITIME RADIONAVIGATION 775A Radiolocation.	MARITIME (80).	

			US44 US316 G56	US44 US316		
3100–3300	RADIOLOCATION. 713 777 778		3100–3300 RADIOLOCATION. 713 778 US110 G59	3100–3300 Radiolocation. 713 778 US110		
3300–3400 RADIOLOCATION 778 779 780	3300–3400 RADIOLOCATION. Amateur. Fixed. Mobile. 778 780	3300–3400 RADIOLOCATION. Amateur. 778 779	3300–3500 RADIOLOCATION. 664 778 US108 G31	3300–3500 Amateur. Radiolocation. 664 778 US108	Amateur (97).	
3400–3600 FIXED. FIXED-SATELLITE (space-to-Earth). Mobile. Radiolocation.	3400–3500 FIXED. FIXED-SATELLITE (space-to-Earth). Amateur. Mobile. Radiolocation 784 664 783					
781 782 785	3500–3700 FIXED. FIXED-SATELLITE (space-to-Earth). MOBILE except aeronautical mobile Radiolocation 784		3500–3600 AERONAUTICAL RADIONAVIGATION (ground-based). RADIOLOCATION US110 G59 G110	3500–3600 Radiolocation. US110		
3600–4200 FIXED. FIXED-SATELLITE (space-to-Earth). Mobile.	786		3600–3700 AERONAUTICAL RADIONAVIGATION (ground-based). RADIOLOCATION US110 US245 G59 G110	3600–3700 FIXED-SATELLITE (space-to-Earth). Radiolocation. US110 US245		
	3700–4200		3700–4200	3700–4200		

International table			United States Table		FCC use designators	
Region 1—allocation MHz	Region 2—allocation MHz	Region 3—allocation MHz	Government	Non-Government	Rule part(s)	Special-use frequencies
			Allocation MHz	Allocation MHz		
(1)	(2)	(3)	(4)	(5)	(6)	(7)
	FIXED. FIXED-SATELLITE (space-to-Earth). MOBILE except aeronautical mobile. 787			FIXED. FIXED-SATELLITE (space-to-Earth). NG41	DOMESTIC PUBLIC FIXED (21). SATELLITE COMMUNICATIONS (25). PRIVATE OPERATIONAL FIXED MICROWAVE (94).	
4200–4400	AERONAUTICAL RADIONAVIGATION 789 788 790 791		4200–4400 AERONAUTICAL RADIONAVIGATION. 791 US2610	4200–4400 AERONAUTICAL RADIONAVIGATION. 791 US261	AVIATION (87).	
4400–4500	FIXED MOBILE.		4400–4500 FIXED. MOBILE.	4400–4500		
4500–4800	FIXED. FIXED-SATELLITE (space-to-Earth). MOBILE. 792A		4500–4800 FIXED. MOBILE. US245	4500–4800 FIXED-SATELLITE (space-to-Earth). 792A US245		
4800–4990	FIXED. MOBILE 793 Radio Astronomy. 720 778 794		4800–4990 FIXED. MOBILE. 720 778 US203 US257	4800–4990 720 778 US203 US257		
4990–5000	FIXED. MOBILE except aeronautical mobile. RADIO ASTRONOMY. Space Research (passive).		4990–5000 RADIO ASTRONOMY Space Research (passive).	4990–5000 RADIO ASTRONOMY Space Research (passive).		

Appendix B

Before the
Federal Communications Commission
Washington, D.C. 20544

In the Matter of

MELBOURNE INTERNATIONAL

COMMUNICATIONS LIMITED

Application for authority to acquire and operate facilities via the PAS satellite system for provision of services between the United States and various overseas points. File No. I-T-C-95–029

ORDER, AUTHORIZATION AND CERTIFICATE

Adopted: February 24, 1995; Released: March 6, 1995

By the Chief, Telecommunications Division:

1. Upon consideration of the above-captioned uncontested application, filed in pursuant to Section 214 of the Communications Act of 1934, as amended, IT IS HEREBY CERTIFIED, that the present and future public convenience and necessity require a grant thereof.

2. Accordingly, IT IS ORDERED, that application File No. I-T-C-95–029 IS GRANTED, and Melbourne International Communications limited (MICL) is authorized to:

a. lease from Pan-American Satellite Corp. (PAS) and operate the satellite circuits specified in the Appendix hereto, between appropriately licensed U.S. earth stations and the PAS-1, PAS-2 and PAS-4 satellites, connecting with similar circuits between the satellites and the overseas points listed in the Appendix hereto, furnished by its correspondents;

b. lease and operate necessary connecting facilities in the United States;

c. lease one-half interest in and operate any necessary overseas connecting facilities; and

d. use the facilities authorized in a, b, and c above to provide MICL's regularly authorized services, including switched voice and data, and private line services, between the United States and the points specified in Appendices A, B, and C.

3. IT IS FURTHER ORDERED that neither the MICL nor any persons or companies directly or indirectly controlling or controlled by MICL, or under direct or indirect common control with it, shall acquire or enjoy any right, for the purpose of handling or interchanging traffic to or from the United States, its territories or possessions, which is denied to any other United States carrier by reason of any concession, contract, understanding, or working arrangement to which MICL or any persons or companies controlling or controlled by MICL are parties.

4. IT IS FURTHER ORDERED that our authorization of MICL to provide private lines as part of its authorized services is limited to the provision of such private lines only between the United States and the countries listed in the Appendices—that is, private lines which originate in the United States and terminate in one of the countries listed in the Appendices or which originate in one of the countries listed in the Appendices and terminate in the United States. In addition, MICL may not — and MICL's tariff must state that its customers may not — connect private lines provided over these facilities to the public switched network at either the U.S., or foreign end, or both, for the provision of international basic switched services, unless authorized to do so by the Commission upon a finding that the destination country affords resale opportunities equivalent to those available under U.S. law, in accordance with *Regulation of International Accounting Rates, Phase II, First Report and Order*, 7 FCC Rcd 559 (1991), Order on Reconsideration and Third Further Notice of Proposed Rulemaking, 7 FCC Rcd 7927 (1994), *petition for reconsideration pending*.

5. IT IS FURTHER ORDERED that the applicant shall file copies of any operating agreements entered into with its foreign correspondents with the Commission within 30 days of their execution, and shall otherwise comply with the filing requirements contained in Section 42.51 of the Commission's Rules, 47 C.F.R. §43.51.

6. IT IS FURTHER ORDERED that the applicant shall file a tariff pursuant to Section 203 of the Communications Act, 47 U.S.C. §203 and Part 61 of the Commission's Rules, 47 C.F.R. Part 61, for the services authorized in this Order.

7. IT IS FURTHER ordered that the applicant shall file the annual reports of overseas telecommunications traffic required by Section 43.61 of the Commission's Rules, 47 C.F.R. §43.61.

8. IT IS FURTHER ORDERED that the applicant shall file a separate Section 214 application for any additional circuits it seeks to,operate via the PAS-1, PAS-2, or PAS-4, satellite.

9. IT IS FURTHER ORDERED that the number of circuits authorized herein for the provision of services via separate satellite systems is subject to limitations on the number of circuits specified under the separate systems policy and applicable consultations under Article XIV(d) of the INTELSAT Agreement.

10. This order is issued under Section 0.261 of the Commission's Rules and is effective upon adoption. Petitions for reconsideration under Section 1.106 or applications for review under Section 1.115 of the Commission's Rules may be filed within 30 days of the date of public notice of this Order (see Section 1.4(b)(2)).

FEDERAL COMMUNICATIONS COMMISSION

Diane J. Cornell
Chief, Telecommunications Division
International Bureau

DA 95-360 **Federal Communications Commission**

APPENDIX

MICL is authorized to lease and operate six 64-kbps voice-grade circuits between the United States and each of the following points, INTERCONNECTED with the public switched network:

PAS-1

C-band countries

Bahamas
Costa Rica
Dominican Republic
Netherlands Antilles
Panama
Peru

Ku-band Countries

Azerbaijan
Bahamas
Czech Republic
Portugal
Romania
Russian Federation
United Kingdom

PAS-2

United Kingdom

PAS-4

Australia
Hong Kong
New Zealand

Glossary

ABC	American Broadcasting Company
AIN	advanced intelligent network
ANTEL	Administración Nacional de Telcomunicaciones
ASME	American Society of Mechanical Engineers
ATIS	Alliance for Telecommunications Industry Solutions
ATM	asynchronous transfer mode
AUSTEL	Australian Telecommunications Authority
BDT	Telecommunications Development Bureau
Bellcore	Bell Communications Research
BfD	federal commissioner for data protection
BOC	Bell operating company
bps	bits per second
BSS	broadcast satellite service
BTO	build, transfer, and operate
CANTV	Compania Anonima Nacional do Telefonos de Venezuela
CAT	Communications Authority of Thailand
CBS	Columbia Broadcasting Company
CCIR	International Radio Consultative Committee
CCIS	common-channel interoffice signaling
CCITT	International Telegraph and Telephone Consultative Committee

CEI	comparably efficient interconnection
CEN	Comite Europeen de Normalisation
CENELEC	Comite Europeen de Normalisation Electrotechnique
CEPT	Conference of European Post and Telecommunications
CEPT	Conference of Post and Telecommunications Administrations
CNT	Comisión Nacional de Telecomunicaciones
CoCom	Coordinating Committee on Multilateral Export Control
CORFO	Corporactión de Fomento
CPE	customer premises equipment
CPT	Compañía Peruana de Teléfonos
CRTC	Canadian Radio-Television and Telecommunications Commission
CSO	community service obligations
DACOM	Data Communications Corporation of Korea
DATE	duly authorized telecommunications entities
DBS	direct broadcast satellite
dc	direct current
DG	directorate general
DIS	draft international standard
DSA	directory service agent
DUA	directory user agent
e-mail	electronic mail
EARC	Extraordinary Administrative Radio Conference
ECSC	European Coal and Steel Community
ECU	European Currency Unit
EEC TREATY	Treaty Establishing the European Economic Community
EFTA	European Free Trade Association
EMETEL	Empresa Estatal de Telcomunicaciones
ENTel	Empresa Nacional de Telecomunicaciones

ENTEL	Empresa Nacional de Telecomunicaciones
ETS	European Telecommunications Standards
ETSI	European Technical Standards Institute
EU	European Union
FCC	Federal Communications Commission
FDM	frequency-division multiplexing
FRR	first-refusal reservation
FSS	fixed-satellite service
GATT	General Agreement on Tariffs and Trade
GDR	German Democratic Republic
GEN	General European Network
GR	guaranteed reservation
GSM	Global System Mobile
GSO	geostationary orbit
HTC	Hungarian Telecommunications Company
Hz	hertz (cycles per second)
I-ETS	Interim European Telecommunications Standards
IBS	international business service
IDDD	international direct distance dialing
IDR	intermediate data rate
IEC	International Electrotechnical Commission
IESS	INTELSAT Earth Station Standards
IETF	Internet Engineering Task Force
IFRB	International Frequency Registration Board
IMTS	international message toll service
INMARSAT	International Maritime Satellite Organization
INTELSAT	International Satellite Communications Organization
IRU	indefeasible right of use
ISDN	integrated services digital network

ISO	International Standards Organization
ISP	International Settlements Policy
ITT	International Telephone and Telegraph Corporation
ITU	International Telecommunications Union
IUC	INTELSAT utilization cost
IVAN	value-added network
Kbps	kilobits per second
KDD	Kokusai Denshin Denwa Company
LAN	local-area network
LATA	local administrative and transport areas
LEO	low Earth orbit
LT	Lithuanian Telecom
M&M	McDonnell & Miller
Mbps	megabits per second
MPT	Ministry of Posts and Telecommunications
ms	millisecond
MSS	mobile-satellite service
MTNL	Mahanagar Telephone Nigam Limited
MTS	message telecommunications service
MUL	multiple-use lease
NAFTA	North American Free Trade Agreement
NBC	National Broadcasting Company
NET	Norme Europeeans des Telecommunications (European Telecommunications Standard)
NFPA	National Fire Protection Association
NPC	North Pacific Cable
NTT	Nippon Telephone and Telegraph Corporation
OECD	Organization for Economic Cooperation and Development
OFAC	Office of Foreign Asset Control

OFTA	Office of the Telecommunications Authority
OFTEL	Office of Telecommunications
ONA	Open Network Architecture
OND	Open Network Development
ONP	Open Network Provision
OSI	Open Systems Interconnection
OTC	Overseas Telecommunications Commission
PAD	packet assembler-disassembler
PanAmSat	PanAmerican Satellite Company Corporation
PAS-3	PanAmSat 3
PBX	private branch exchange
PCA	protective coupling arrangement
PCM	pulse-code modulator
PCS	personal communications service
PLDT	Philippine Long-Distance Telephone Company
PT Telkom	PT Telekomunikasi Indonesia
PTC	Pakistan Telecommunications Corporation
PTT	postal, telegraph, and telephone
RARC	Regional Administrative Radio Conference
RF	radio frequency
RHC	regional holding companies
RRB	Radio Regulations Board
SBS	Satellite Business Systems
SC	subcommittee
SCT	Secretariat for Communications and Transportation
SDN	software-defined network
SMDS	switched multimegabit data service
SOG-T	Senior Officials Group on Telecommunication
SONET	Synchronous Optical Network

ST	Singapore Telecom
STA	special temporary authority
TC	technical committee
TCNZ	Telecom Corporation of New Zealand
TCP/IP	Transport Control Protocol/Internet Protocol
TDM	time-division multiplexing
TELECOM	Empresa Nacional de Telecomunicaciones
TelMex	Telefonos Mexico
TEN	Telecommunications European Networks
TOT	Telephone Organization of Thailand
TPSA	Telekomunikacja Polska
TRAC	Technical Recommendations Application Committee
TRI	Technologies Resource Industries
UN	United Nations
USP	Uniform Settlements Policy
USTR	United States Trade Representative
VSNL	Videsh Sanchar Nigam Limited
WARC	World Administrative Radio Conference
WARC-ST	World Administrative Radio Conference for Space Telecommunications
WATS	Wide Area Telecommunications Service
WRC	World Radiocommunication Conference

About the Authors

Charles H. Kennedy is a telecommunications attorney practicing with Morrison & Foerster in Washington, DC. He was a member of the defense trial team in a number of the antitrust cases against the Bell System and served more recently as an antitrust and regulatory attorney for a Bell operating company. Mr. Kennedy is a 1976 graduate of the University of Chicago Law School, where he was an editor of *The University of Chicago Law Review.* He is the author of *An Introduction to U.S. Telecommunications Law*, also published by Artech House.

M. Veronica Pastor is a telecommunications attorney practicing with the firm of Squire, Sanders & Dempsey in Washington, D.C. She has represented clients before the FCC and INTELSAT. Ms. Pastor is a 1992 graduate of Georgetown University Law Center.

Index

The Artech House Telecommunications Library

Vinton G. Cerf, Series Editor

Advanced Technology for Road Transport: IVHS and ATT, Ian Catling, editor

Advances in Computer Communications and Networking, Wesley W. Chu, editor

Advances in Computer Systems Security, Rein Turn, editor

Advances in Telecommunications Networks, William S. Lee and Derrick C. Brown

Analysis and Synthesis of Logic Systems, Daniel Mange

An Introduction to International Telecommunications Law, Charles H. Kennedy and M. Veronica Pastor

An Introduction to U.S. Telecommunications Law, Charles H. Kennedy

Asynchronous Transfer Mode Networks: Performance Issues, Raif O. Onvural

ATM Switching Systems, Thomas M. Chen and Stephen S. Liu

A Bibliography of Telecommunications and Socio-Economic Development, Heather E. Hudson

Broadband: Business Services, Technologies, and Strategic Impact, David Wright

Broadband Network Analysis and Design, Daniel Minoli

Broadband Telecommunications Technology, Byeong Lee, Minho Kang, and Jonghee Lee

Cellular Radio: Analog and Digital Systems, Asha Mehrotra

Cellular Radio Systems, D. M. Balston and R. C. V. Macario, editors

Client/Server Computing: Architecture, Applications, and Distributed Systems Management, Bruce Elbert and Bobby Martyna

Codes for Error Control and Synchronization, Djimitri Wiggert

Communications Directory, Manus Egan, editor

The Complete Guide to Buying a Telephone System, Paul Daubitz

Computer Telephone Integration, Rob Walters

The Corporate Cabling Guide, Mark W. McElroy

Corporate Networks: The Strategic Use of Telecommunications, Thomas Valovic

Current Advances in LANs, MANs, and ISDN, B. G. Kim, editor

Digital Cellular Radio, George Calhoun

Digital Hardware Testing: Transistor-Level Fault Modeling and Testing, Rochit Rajsuman, editor

Digital Signal Processing, Murat Kunt

Digital Switching Control Architectures, Giuseppe Fantauzzi

Distributed Multimedia Through Broadband Communications Services, Daniel Minoli and Robert Keinath

Disaster Recovery Planning for Telecommunications, Leo A. Wrobel

Document Imaging Systems: Technology and Applications, Nathan J. Muller

EDI Security, Control, and Audit, Albert J. Marcella and Sally Chen

Electronic Mail, Jacob Palme

Enterprise Networking: Fractional T1 to SONET, Frame Relay to BISDN, Daniel Minoli

Expert Systems Applications in Integrated Network Management, E. C. Ericson, L. T. Ericson, and D. Minoli, editors

FAX: Digital Facsimile Technology and Applications, Second Edition, Dennis Bodson, Kenneth McConnell, and Richard Schaphorst

FDDI and FDDI-II: Architecture, Protocols, and Performance, Bernhard Albert and Anura P. Jayasumana

Fiber Network Service Survivability, Tsong-Ho Wu

Fiber Optics and CATV Business Strategy, Robert K. Yates *et al.*

A Guide to Fractional T1, J. E. Trulove

A Guide to the TCP/IP Protocol Suite, Floyd Wilder

Implementing EDI, Mike Hendry

Implementing X.400 and X.500: The PP and QUIPU Systems, Steve Kille

Inbound Call Centers: Design, Implementation, and Management, Robert A. Gable

Information Superhighways: The Economics of Advanced Public Communication Networks, Bruce Egan

Integrated Broadband Networks, Amit Bhargava

Intelcom '94: The Outlook for Mediterranean Communications, Stephen McClelland, editor

International Telecommunications Management, Bruce R. Elbert

International Telecommunication Standards Organizations, Andrew Macpherson

Internetworking LANs: Operation, Design, and Management, Robert Davidson and Nathan Muller

Introduction to Document Image Processing Techniques, Ronald G. Matteson

Introduction to Error-Correcting Codes, Michael Purser

Introduction to Satellite Communication, Bruce R. Elbert

Introduction to T1/T3 Networking, Regis J. (Bud) Bates

Introduction to Telecommunication Electronics, Second Edition, A. Michael Noll

Introduction to Telephones and Telephone Systems, Second Edition, A. Michael Noll

Introduction to X.400, Cemil Betanov

Land-Mobile Radio System Engineering, Garry C. Hess

LAN/WAN Optimization Techniques, Harrell Van Norman

LANs to WANs: Network Management in the 1990s, Nathan J. Muller and Robert P. Davidson

Long Distance Services: A Buyer's Guide, Daniel D. Briere

Measurement of Optical Fibers and Devices, G. Cancellieri and U. Ravaioli

Meteor Burst Communication, Jacob Z. Schanker

Minimum Risk Strategy for Acquiring Communications Equipment and Services, Nathan J. Muller

Mobile Communications in the U.S. and Europe: Regulation, Technology, and Markets, Michael Paetsch

Mobile Information Systems, John Walker

Narrowband Land-Mobile Radio Networks, Jean-Paul Linnartz

Networking Strategies for Information Technology, Bruce Elbert

Numerical Analysis of Linear Networks and Systems, Hermann Kremer *et al.*

Optimization of Digital Transmission Systems, K. Trondle and Gunter Soder

Packet Switching Evolution from Narrowband to Broadband ISDN, M. Smouts

Packet Video: Modeling and Signal Processing, Naohisa Ohta

Personal Communication Systems and Technologies, John Gardiner and Barry West, editors

The PP and QUIPU Implementation of X.400 and X.500, Stephen Kille

Practical Computer Network Security, Mike Hendry

Principles of Secure Communication Systems, Second Edition, Don J. Torrieri

Principles of Signaling for Cell Relay and Frame Relay, Daniel Minoli and George Dobrowski

Principles of Signals and Systems: Deterministic Signals, B. Picinbono

Private Telecommunication Networks, Bruce Elbert

Radio-Relay Systems, Anton A. Huurdeman

Radiodetermination Satellite Services and Standards, Martin Rothblatt

Residential Fiber Optic Networks: An Engineering and Economic Analysis, David Reed

Secure Data Networking, Michael Purser

Service Management in Computing and Telecommunications, Richard Hallows

Setting Global Telecommunication Standards: The Stakes, The Players, and The Process, Gerd Wallenstein

Smart Cards, José Manuel Otón and José Luis Zoreda

Super-High-Definition Images: Beyond HDTV, Naohisa Ohta, Sadayasu Ono, and Tomonori Aoyama

Television Technology: Fundamentals and Future Prospects, A. Michael Noll

Telecommunications Technology Handbook, Daniel Minoli

Telecommuting, Osman Eldib and Daniel Minoli

Telemetry Systems Design, Frank Carden

Telephone Company and Cable Television Competition, Stuart N. Brotman

Teletraffic Technologies in ATM Networks, Hiroshi Saito

Terrestrial Digital Microwave Communications, Ferdo Ivanek, editor

Toll-Free Services: A Complete Guide to Design, Implementation, and Management, Robert A. Gable

Transmission Networking: SONET and the SDH, Mike Sexton and Andy Reid

Transmission Performance of Evolving Telecommunications Networks, John Gruber and Godfrey Williams

Troposcatter Radio Links, G. Roda

Understanding Emerging Network Services, Pricing, and Regulation, Leo A. Wrobel and Eddie M. Pope

UNIX Internetworking, Uday O. Pabrai

Virtual Networks: A Buyer's Guide, Daniel D. Briere

Voice Processing, Second Edition, Walt Tetschner

Voice Teletraffic System Engineering, James R. Boucher

Wireless Access and the Local Telephone Network, George Calhoun

Wireless Data Networking, Nathan J. Muller

Wireless LAN Systems, A. Santamaría and F. J. López-Hernández

Wireless: The Revolution in Personal Telecommunications, Ira Brodsky

Writing Disaster Recovery Plans for Telecommunications Networks and LANs, Leo A. Wrobel

X Window System User's Guide, Uday O. Pabrai

For further information on these and other Artech House titles, contact:

Artech House
685 Canton Street
Norwood, MA 02062
617-769-9750
Fax: 617-769-6334
Telex: 951-659
email: artech@world.std.com

Artech House
Portland House, Stag Place
London SW1E 5XA England
+44 (0) 171-973-8077
Fax: +44 (0) 171-630-0166
Telex: 951-659
email: bookco@artech.demon.co.uk